The Index Theorem and the Heat Equation

Peter B. Gilkey

Princeton University

Notes by Jon Sacks

ISBN 0-914098-13-6
Library of Congress Catalog Card Number: 74-84575

PUBLISH OR PERISH, INC.
6 BEACON STREET
BOSTON, MASS. 02108 (U.S.A.)

In Japan distributed exclusively by
KINOKUNIYA BOOK-STORE CO., LTD.
TOKYO, JAPAN

INTRODUCTION

This is a set of notes which was taken by Jon Sacks in a seminar which I gave on the index theorem during the winter and spring semesters of 1973 at Berkeley. The seminar dealt with the proof of the index theorem for the DeRham, Dolbeault, and Signature complexes using heat equation methods for manifolds without boundary. The calculus of pseudo-differential operators which is necessary is developed in detail in the first three chapters. The fourth chapter gives the proof in detail of the classical Chern-Gauss-Bonnet formula for the Euler characteristic of a manifold without boundary. The fifth and sixth chapters sketch the proof of the corresponding results for the Dolbeault and Signature complexes. The proof of these cases is similar to the proof for the DeRham complex and is therefore only sketched briefly.

I would like to express my gratitude to Jon Sacks for writing up these notes--he hadan often difficult task of making my oral arguments into readable mathematics. I would also like to thank my advisor , Professor L. Nirenberg, from whose lectures at NYU on pseudo-differential operators much of the first three chapters is taken. Finally, I would like to thank Professor Chern for all his help and encouragement while I was at Berkeley.

Peter B. Gilkey
June 15,1974

PAGE CONTENTS

ELLIPTIC OPERATORS ON COMPACT MANIFOLDS

Peter Gilkey, Winter, 1974

Chapter I. Pseudo-differential operators.

In our discussion of elliptic operators we will be using the calculus of pseudo-differential operators (ψ DO's). First we have to find a suitable class of functions for these to act on. This class turns out to be the family of Sobolev spaces H_s, which are the completion of the Schwarz class $\mathcal{S}$ in a suitable norm. First we work on Euclidean space R^N. All functions on R^N will be complex-valued. Greek letters will denote multi-indices if they are near the beginning of the alphabet

e.g. $\alpha = (\alpha_1,\ldots,\alpha_N) \in (\mathbf{Z}^+)^N$

$$|\alpha| = \sum_{i=1}^{N} \alpha_i \qquad \alpha! = \alpha_1!\alpha_2! \ldots\alpha_N!$$

$$D_x^\alpha = (-i)^{|\alpha|}\frac{\partial^{\alpha_1}}{\partial x_1^{\alpha_1}} \cdots \frac{\partial^{\alpha_N}}{\partial x_N^{\alpha_N}}$$

$$x^\alpha = x_1^{\alpha_1}\ldots x_N^{\alpha_N} \qquad \xi^\alpha = \xi_1^{\alpha_1}\ldots\xi_N^{\alpha_N} \quad \text{for } x,\ \xi \in R^N.$$

Pairs (x,ξ) should be thought of as elements of $T^*(R^N)$, the cotangent bundle of R^N..i.e. ξ is a covector on R^N. All integrals will be with respect to the measure

$(2\pi)^{-N/2}x$ (Lebesgue measure on R^N).

$C_0^\infty(R^N)$ denotes smooth functions with compact support.

<u>The Fourier Transform on the Schwarz class $\mathcal{S}(R^N)$.</u>

$\mathcal{S}(R^N) = \{u \in C^\infty(R^N) : \forall(\alpha,n)\ \exists$ constant $C_{\alpha,n}$ with $|D^\alpha u| \leqslant C_{\alpha,n}(1+|x|)^{-n}\}$. The functions in $\mathcal{S}$ have all their derivatives decreasing faster at infinity than the inverse of any polynomial. It can be shown that $C_0^\infty(R^N) \subset \mathcal{S}(R^N) \subset L^2(R^N)$. Since C_0^∞ is dense in L^2, it follows that $\mathcal{S}$ is dense in L^2. We have the classical Fourier Transform $\mathcal{F}:\mathcal{S} \to \mathcal{S}$ homeomorphically, $\mathcal{F}(u)\ (\xi) = \hat{u}(\xi) = \int_{R^N} e^{-ix\cdot\xi}u(\xi)d\xi$.

Let (u,v) denote $\int u\bar{v}$ and $(u*v)(x) = \int u(x-y)v(y)dy$. Then $\mathcal{F}$ has the following properties:

(1) $(u,v) = (\hat{u},\hat{v})$ (Parceval's formula)

(2) $\widehat{u*v} = \hat{u}\hat{v}$

$\widehat{u\cdot v} = \hat{u}*\hat{v}$

(3) $u(x) = \int e^{ix\cdot\xi}\hat{u}(\xi)d\xi$ (Inversion formula)

(4) $\widehat{D^\alpha u}(\xi) = \xi^\alpha \hat{u}(\xi)$

(5) $\widehat{X^\beta u}(\xi) = D_\xi^\beta \hat{u}(\xi)$

(1) and (3) and the fact that $\mathcal{S}$ is dense in L^2 lead to the <u>Plancherel Theorem</u>: There is a unique continuous extension of $\mathcal{F}$ to an isomorphism of L^2 with itself and with this extension (1) holds for $u,v \in L^2$.

(4) and (5) are the main justification for $\mathcal{F}$, as they enable us to replace differentiation by multiplication and vice versa by going to the transform space.

<u>The Sobolev spaces $H_s(R^N)$.</u>

For s a real number and $u \in \mathcal{S}$ define

$$\|u\|_s = (\int|\hat{u}(\xi)|^2(1+|\xi|^2)^s d\xi)^{1/2}$$

The Sobolev space $H_s(R^N)$ is defined to be the completion of $\mathcal{S}(R^N)$ in the norm $\|\ldots\|_s$.

H_s corresponds to L^2 with the weighted measure $(1+|\xi|^2)^s d\xi$. Replacing $(1+|\xi|^2)^s$ by $(1+|\xi|)^{2s}$ in the definition of $\|\ldots\|_s$ gives rise to an equivalent norm since there are constants c_1 and c_2 such that $c_1(1+|\xi|^2)^s \leq (1+|\xi|)^{2s} \leq c_2(1+|\xi|^2)^s$. With this equivalent norm, $\|u\|_s$ is just the L^2 norm of $(1+|\xi|)^s\hat{u}(\xi)$. Recalling that $\mathcal{F}$ is an isometry on L^2 and that multiplication comes from differentiation, we see that in some sense s "counts" the number of L^2 derivatives of elements in H_s. This is precisely what happens when s is a positive integer. In fact for $n \in \mathbb{Z}^+$ we could define

$$\|u\|_n = \sum_{|\alpha|\leq n} \int|\xi^\alpha\hat{u}(\xi)|^2 d\xi$$

$$= \int\{\sum_{|\alpha|\leq n} |\xi^\alpha|\}^2|\hat{u}(\xi)|^2 d\xi$$

Under the inverse Fourier transform this last expression corresponds to

$$\int \Big\{ \sum_{|\alpha| \leq n} |D^{\alpha}u(x)|^2 \Big\} dx$$

Now there are constants k_1 and k_2 such that

$$k_1(1+|\xi|)^{2n} \leq \Big(\sum_{|\alpha| \leq n} |\xi^{\alpha}| \Big)^2 \leq k_2(1+|\xi|)^{2n}$$

so that $\|u\|_n$ is equivalent to our original definition. With the above interpretation in mind it is not surprising that when we extend the operator D^{α} from $\mathcal{S}$ to H_s we get the following

<u>Lemma</u>: $D_x^{\alpha} : H_s \to H_{s-|\alpha|}$ is a continuous operator.

<u>Proof</u>. We need the estimate $|\xi^{\alpha}|^2(1+|\xi|^2)^{s-|\alpha|} \leq C(1+|\xi|^2)^s$.

$$\|D_x^{\alpha}u\|_{s-|\alpha|}^2 = (\int (|\xi^{\alpha}||\hat{u}(\xi)|)^2(1+|\xi|^2)^{s-|\alpha|}d\xi)$$

$$\leq C\|u\|_s$$

<u>Theorem</u> (Sobolev). Let $s > N/2$. Then

(1) $u \in H_s \Rightarrow u \in C^0(R^N)$

(2) $u \in H_{s+k} \Rightarrow u \in C^k(R^N)$.

Proof (1). Let $u \in \mathcal{S}$. Then $u(x) = \int e^{ix\cdot\xi}\hat{u}(\xi)d\xi$

$$= \int e^{ix\cdot\xi}(1+|\xi|^2)^{-s/2}(1+|\xi|^2)^{s/2}\hat{u}(\xi)d\xi.$$

Therefore $|u(x)| \leq (\int(1+|\xi|^2)^{-s}d\xi)^{1/2}(\int|\hat{u}(\xi)|^2(1+|\xi|^2)^s d\xi)^{1/2}$ by Schwarz's inequality.

But $\int(1+|\xi|^2)^{-s}d\xi = c_s < \infty$ since $s > N/2$ so $|u(x)| \leq c_s\|u\|_s$ uniformly in x.

Thus convergence in $\|\ldots\|_s$ implies uniform convergence. Now, if $u \in H_s$, it is the limit in $\|\ldots\|_s$ of a sequence from $\mathcal{S}$, thus the uniform limit of this sequence, hence continuous.

(2) Let $u \in H_{s+k}$. The lemma implies $D_x^\alpha u \in H_s$ for $|\alpha| \leq k$ and the result follows from (1)

Observe that $s \geq t \Rightarrow (1+|\xi|^2)^s \geq (1+|\xi|^2)^t \Rightarrow H_s \subset H_t$.

The next result tells us that in some sense this inclusion is compact, an important fact which will be very useful later on.

Theorem. Let $\{u_n\} \subset H_s$ be such that supp $u_n \subset C$ (a compact set) and that $\|u_n\|_s \leq 1$ for all n. Then $\{u_n\}_{n=1}^\infty$ contains a subsequence converging in H_t if $t < s$.

<u>Proof</u>: We first show that $\{\hat{u}_n\}$ converges uniformly on compact subsets.

Choose $\phi \in C_0^\infty(R^N)$ such that $\phi \equiv 1$ on $\bigcup_n \operatorname{supp} u_n$

$$\widehat{\phi u_n} = \hat{u}_n = \hat{\phi} * \hat{u}_n \quad \text{so that}$$

$$D_j\hat{u}_n = D_j\hat{\phi} * \hat{u}_n \quad (D_j = (-i)\frac{\partial}{\partial \xi^j}).$$

Thus $|D_j \hat{u}_n(\xi)| = |\int D_j\hat{\phi}(\xi-\eta)\hat{u}_n(\eta)d\eta|$. By Schwartz inequality,

$\leqslant [\int(1+|\eta|^2)^{-s}|D_j\hat{\phi}(\xi-\eta)|^2 d\eta]^{1/2}[\int|u_n(\eta)|^2(1+|\eta|^2)^s d\eta]^{1/2}$
$\leqslant f(\xi)\|u_n\|_s \leqslant f(\xi)$.

Therefore for each j, the $\{D_j\hat{u}_n\}$ are uniformly bounded on compact ξ-subsets as are $\{\hat{u}_n\}$.
It follows that the $\{\hat{u}_n\}$ are uniformly equicontinuous on compact ξ-subsets. By the Arzelá-Ascoli theorem $\{\hat{u}_n\}$ contains a subsequence, which we also label $\{\hat{u}_n\}$, which converges uniformly on compact ξ-subsets. To prove the theorem it is sufficient to prove that $\{u_n\}$ is Cauchy in H_t

$$\|u_n-u_m\|_t^2 = \int_{R^n}|\hat{u}_n-\hat{u}_m|^2(1+|\xi|^2)^t d\xi = \int_{|\xi|\leqslant r} + \int_{|\xi|>r}$$

Now $\int_{|\xi|>r}|\hat{u}_n-\hat{u}_m|^2(1+|\xi|^2)^t d\xi \leqslant \dfrac{1}{(1+r^2)^{s-t}}\int_{|\xi|>r}|\hat{u}_n-\hat{u}_m|^2(1+|\xi|^2)^s d\xi$

Since $(s-t)>0 \Rightarrow (1+|\xi|^2)^t \leqslant \dfrac{1}{(1+r^2)^{s-t}}(1+|\xi|^2)^s$ for $|\xi| > r$

Therefore $\int_{|\xi|\geq r}|\hat{u}_n-\hat{u}_m|^2(1+|\xi|^2)^t d\xi \leq \frac{1}{(1+r^2)^{s-t}}\|u_n-u_m\|_s^2 \leq \frac{4}{(1+r^2)^{s-t}}$

On the other hand $\int_{|\xi|\leq r}|\hat{u}_n(\xi)-\hat{u}_m(\xi)|^2(1+|\xi|^2)^t d\xi \to 0$ since $\{u_n\}$ converges uniformly on compact ξ-subsets. Thus for n,m and r large enough, $\|u_n-u_m\|_t^2 < \varepsilon$. Since H_t is complete, $\{u_n\}$ converges in H_t.

The hypothesis in the above theorem that $\{\text{supp } u_n\} \subset C$ for C compact is essential. It is easy to construct a sequence $\{u_n\}$ such that $\|u_n\|_s = 1$ for all n and so the supports are pair-wise disjoint. In this case we can find $\varepsilon > 0$ so that $\|u_n-u_m\|_t > \varepsilon$ however large n,m are.

Each H_s, being a Hilbert space, is its own dual, but since there is no preferred norm on H_s, it is more useful to consider the following characterization of $(H_s)^*$.

<u>Proposition</u>: The pairing of $H_s \times H_{-s} \to \mathbb{C}$ defined by

$$(u,v) \to \int \hat{u}\hat{v}d\xi$$

is a perfect pairing, and so identities $(H_s)^*$ with H_{-s} and $(H_{-s})^*$ with H_s.

<u>Proof</u>: Denote $\int \hat{u}\hat{v}d\xi$ be $\langle u, v\rangle$. Actually, one first defines $\langle u,v\rangle$ for $u,v \in \mathcal{S}$ and then extends the definition by continuity to $u,v \in H_s \times H_{-s}$. One checks easily that $\langle u,v\rangle$ is then well-defined.

Now $|\langle u,v\rangle| = |\langle \hat{u},\hat{v}\rangle| = |\int \hat{u}\hat{v}d\xi|$

$$= |\int \hat{u}(\xi)(1+|\xi|^2)^{s/2}\ \hat{v}(\xi)(1+|\xi|^2)^{-s/2}d\xi|$$

$$\leq [\int |\hat{u}(\xi)|^2(1+|\xi|^2)^s d\xi]^{1/2}[\int |\hat{v}(\xi)|^2(1+|\xi|^2)^{-s}d\xi]^{1/2}$$

$$= \|u\|_s\ \|v\|_{-s}\ .$$

Thus for fixed $u \in H_s$, $\sup\limits_{v\in H_{-s}} \dfrac{|\langle u,v\rangle|}{\|v\|_{-s}} \leq \|u\|_s$

which shows that u defines a bounded linear functional on H_{-s}. If we can show that the sup on the left-hand side is actually attained, we will have an isometric embedding of H_s in $(H_{-s})^*$. To do this choose $v_0 \in H_{-s}$ such that

$$\hat{v}_0(\xi) = \hat{u}(\xi)(1+|\xi|^2)^s.$$

Then $|\langle u,v_0\rangle| = \int |\hat{u}(\xi)|^2(1+|\xi|^2)^s d\xi = \|u\|_s^2.$

But $\|v_0\|_{-s} = \int |\hat{u}(\xi)|^2(1+|\xi|^2)^{2s}(1+|\xi|^2)^{-s}d\xi$

$$= \int |\hat{u}(\xi)|^2(1+|\xi|^2)^s d\xi = \|u\|_s.$$

Thus $\dfrac{|\langle u,v_0\rangle|}{\|v_0\|_{-s}} = \dfrac{\|u\|_s^2}{\|u\|_s} = \|u\|_s.$

Similarly, $H_{-s} = (H_s)^*$. So the pairing $(u,v) \mapsto \langle u,v\rangle$ is a perfect pairing and H_s and H_{-s} are dual to each other.

Alternately, since $H_s = L^2$ with weight function $(1+|\xi|^2)^s$
and $H_{-s} = L^2$ with weight function $(1+|\xi|^2)^{-s}$
and L^2 is canonically self-dual, we get H_s and H_{-s} dual to each other.

Pseudo-differential operators on H_s.

We are now in a position to define what a pseudo-differential operator is. First recall that a linear differential operator is a polynomial expression

$$P(x,D) = \sum_{|\alpha|\leq m} a_\alpha(x) D_x^\alpha$$

The symbol of P, denoted $\sigma P = p(x,\xi) = \sum_{|\alpha|\leq m} a_\alpha(x)\xi^\alpha$. Observe that $D_x^\alpha u(\xi) = \xi^\alpha \hat{u}(\xi)$, and taking $\mathcal{F}^{-1}$ of both sides we get

$$D_x^\alpha u(x) = \int e^{ix\cdot\xi}\, \xi^\alpha \hat{u}(\xi)d\xi \ .$$

Similarly, $P(x,D)u(x) = \int e^{ix\cdot\xi} p(x,\xi)\hat{u}(\xi)d\xi$.

We use this last formula to define the action of a pseudo-differential operator for a wider class of functions $p(x,\xi)$ than polynomials.

Definition: $p(x,\xi)$ is said to be a symbol of order m, denoted $p \in S^m$, if it satisfies

(1) $p(x,\xi)$ is C^∞ in x and ξ

(2) $p(x,\xi)$ has compact x-support

(3) for all α,β, there is a constant $C_{\alpha\beta}$ s.t. $|D_x^\alpha D_\xi^\beta p(x,\xi)| \leq C_{\alpha\beta}(1+|\xi|)^{m-|\beta|}$

Let $P(x,D) = \sum_{|\alpha|\leq m} a_\alpha(x)D^\alpha$ be a differential operator. The symbol of P, $\sigma(P) = = \sum_{|\alpha|\leq m} \xi^\alpha a_\alpha(x)$ is a polynomial of order m. Consequently it lies in S^m and therefore P is a pseudo-differential operator of order m. We proved earlier that $P: H_s \dashrightarrow H_{s-m}$ is a continuous linear operator. We will now extend this theorem to pseudo-differential operators.

<u>Theorem</u>: Let $\sigma(P) = p \in S^m$. Then $P:H_s \to H_{s-m}$ is a continuous linear operator for all s.

<u>Proof</u>: We compute:

$$Pu(x) = \int e^{ix\cdot\xi} p(x,\xi)\hat{u}(\xi)d\xi.$$

Therefore $\widehat{Pu}(\eta) = \iint e^{ix\cdot(\xi-\eta)} p(x,\xi)\hat{u}(\xi)d\xi dx$

$$= \iint e^{ix\cdot(\xi-\eta)} p(x,\xi)\hat{u}(\xi)dxd\xi .$$

Define $\hat{P}_x(\lambda,\xi) = \int e^{-ix\cdot\lambda} p(x,\xi)dx.$

Then $\widehat{Pu}(\eta) = \int \hat{P}_x(\eta-\xi,\xi)\hat{u}(\xi)d\xi.$

Now let $\phi \in H_{m-s}$. Then

$$\int \widehat{Pu}(\eta)\hat{\phi}(\eta)d\eta = \iint \hat{P}_x(\eta-\xi,\xi)\hat{\phi}(\eta)\hat{u}(\xi)d\xi d\eta$$

$$= \iint P_x(\eta-\xi,\xi)(1+|\xi|)^{-s}(1+|\eta|)^{s-m}\hat{\phi}(\eta)(1+|\eta|)^{m-s}\hat{u}(\xi)(1+|\xi|)^s d\xi d\eta$$

If we can show that $K(\xi,\eta) = |\hat{P}_x(\eta-\xi,\xi)(1+|\xi|)^{-s}(1+|\eta|)^{s-m}|$ is such that

$$\int |K(\xi,\eta)|d\xi \leq C \quad \text{and} \quad \int |K(\xi,\eta)|d\eta \leq C .$$

Then:

$$|\int Pu(\eta)\hat{\phi}(\eta)d\eta| \leqslant |\iint K(\xi,\eta)(1+|\eta|)^{m-s}\hat{\phi}(\eta)(1+|\xi|)^{s}\hat{u}(\xi)d\xi d\eta|$$

$$\leqslant C(\iint K(\xi,\eta)[(1+|\xi|)^{s}\hat{u}(\xi)]^{2}d\xi d\eta)^{1/2}(\iint K(\xi,\eta)((1+|\eta|)^{m-s}\hat{\phi}(\eta))^{2}d\xi d\eta)$$

$$\leqslant C\|\phi\|_{m-s}\|u\|_{s} .$$

Since H_{m-s} is dual to H_{s-m}, this implies $\|Pu\|_{s-m} \leqslant C\|u\|_{s}$ and therefore P is a bounded operator from $H_s \to H_{s-m}$.

We prove $\int|K(\xi,\eta)|d\xi \leqslant C$ and $\int|K(\xi,\eta)|d\eta \leqslant C$ as follows: $p \in S^m$ implies $|D^{\alpha}_{x}p| \leqslant C(1+|\xi|)^m$. Since p has compact x-support,

$$\Rightarrow |D^{\alpha}_{x}p(\lambda,\xi) = |\int e^{-i\,\cdot\lambda}D^{\alpha}_{x}p(x,\xi)dx| \leqslant C(1+|\xi|\ )^m$$

$$\Rightarrow |D^{\alpha}_{x}p(\lambda,\xi)| \leqslant C(1+|\xi|)^m$$

$$\Rightarrow |D^{\alpha}_{x}p(\lambda,\xi)| \leqslant C(1+|\xi|)^m(1+|\lambda|)^{-k}.$$

C is a generic constant which is different in each equation.

$$\Rightarrow |K| = |\hat{P}_x(\eta-\xi,\eta)(1+|\eta|)^{-s}(1+|\eta|)^{s-m}|$$

$$\leqslant C\,\frac{(1+|\xi|)^{m-s}}{(1+|\eta|)^{m-s}}\,(1+|\xi-\eta|)^{-k}$$

the inequality: $\frac{1+|a|}{1+|b|} \leqslant 1+|a-b|$ implies $|K| \leqslant C(1+|\xi-\eta|)^{-k+|m-s|}$

therefore $\int|k(\xi,\eta)d\xi| \leqslant \int C(1+|\xi-\eta|)^{-k+|m-s|}d\xi < +\infty$

$$\int|k(\xi,\eta)d\eta| \leqslant \int C(1+|\xi-\eta|)^{-k+|m-r|}d\eta < +\infty$$

if k is large. Q.E.D.

Examples of symbols of order m.

1) $p(x,\xi) = \sum_{|\alpha|\leq m} C_\alpha(x)\xi^\alpha \in S^m$ where the $C_\alpha(x)$ are C_0^∞ functions

all of whose supports are contained in some compact set K. P is a polynomial of order m and the associated operator $P = \sum_{|\alpha|\leq m} C_\alpha(x)D^\alpha$ is a differential operator

2) $p(x,\xi) = c(x)(1+|\xi|^2)^{m/2} \in S^m$, where $c(x) \in C_0^\infty$ and m is not necessarily an integer. If $m = -2$, then

$p(x,\xi) = \dfrac{c(x)}{1+|\xi|^2}$ and $P:H_s \to H_{s+2}$ for all s, so that P makes

functions smoother, which is not the case with differential operators.

Our next task is to show that the class of ψDO's forms an algebra closed under the operations of composition of operators and taking the adjoint of an operator. Before we do that, we see what happens with some easy examples of the simpler class of differential operators:

Consider the following operators on $\mathcal{S}(\mathbb{R})$

$$P = \frac{1}{i}\, a(x)\frac{\partial}{\partial x} \qquad Q = b(x) \quad \text{where} \quad a(x),\, b(x) \in C_0^\infty(\mathbb{R})$$

$$\sigma P = a(x)\xi \qquad \sigma Q = b(x).$$

For $u \in \mathcal{S}(R)$, $(PQ)u = P(Qu) = \frac{1}{i}a(x)\frac{\partial}{\partial x}\,(b(x)u(x))$

$$= \frac{1}{i}a(x)[b'(x)u(x)+b(x)u'(x)] = (\frac{1}{i}ab\frac{\partial}{\partial x} + \frac{1}{i}ab')u.$$

Thus $PQ = \frac{1}{i}ab\frac{\partial}{\partial x} + \frac{1}{i}\,ab'$ so that $\sigma(PQ) = ab\xi + \frac{1}{i}ab'$

$$= \sigma P\cdot\sigma Q + \partial_\xi^1\sigma P\cdot D_x^1\sigma Q$$

$$= \sum_{\alpha=0,1} \partial_\xi^\alpha\sigma P D_x^\alpha\sigma Q$$

Let P^* denote the adjoint of P with respect to the L^2 inner product i.e. $(P^*u,v) = (u,Pv)$. We compute P^* for $P = \frac{1}{i}a(x)\frac{\partial}{\partial x}$ as follows:

$$\int \frac{1}{i}a(x)\frac{\partial}{\partial x}{}^{*}u\bar{v}dx = \int u\,\overline{\frac{1}{i}a(x)\frac{\partial}{\partial x} v}dx$$

$$= i\int u\bar{a}\bar{v}'dx = i[\bar{a}u\bar{v}\Big|_{-\infty}^{\infty} - \int(\bar{a}u)'\bar{v}dx], \quad \text{integrating by parts.}$$

But $\bar{a}u\bar{v}\Big|_{-\infty}^{\infty} = 0$ since a has compact support.

Therefore $(\frac{1}{i}a\frac{\partial}{\partial x})^*u = \frac{1}{i}(\bar{a}u)'$ so that $(\frac{1}{i}a\frac{\partial}{\partial x})^* = \frac{1}{i}\bar{a}\frac{\partial}{\partial x} + \frac{1}{i}(\bar{a}\ )'$ so that

$(\frac{1}{i}a\frac{\partial}{\partial x})^* = \frac{1}{i}a^*\frac{\partial}{\partial x} + \frac{1}{i}(a^*)'$, setting $\bar{a} = a^*$.

Thus $\sigma(P^*) = (\sigma(P))^* + \partial_\xi^1 D_x^1(\sigma(P))^* = \sum_{\alpha=0,1} \partial_\xi^\alpha D_x(\sigma(P))^*.$

We shall get similar formulas for $\sigma(PQ)$ and $\sigma(P^*)$ when P and Q are ψDO's except that the sums involved will be infinite rather than finite.

<u>The Algebra of Pseudo-differential operators.</u>

Henceforth the corresponding capital letter of a small letter will denote the associated operator of a symbol. We introduce an equivalence relation on the class of symbols by defining $a \sim b \Leftrightarrow a - b \in S^m$ for all m. Thus $A\text{-}B: H_s \to H_{s'}$ for all s, s', if $a \sim b$, which implies $A\text{-}B: H_s \to C^\infty$ for all s by Sobolev's theorem. Thus we quotient out by the class of infinitely smoothing operators, whose effect we can ignore from our point of view.

<u>Definition</u>: Given symbols a, $\{a_j\}_{j=0}^\infty$ we write

$$a \sim \sum_{j=0}^{\infty} a_j \quad \text{provided } \forall m \in \mathbb{R}^+ \ \exists N \in \mathbb{Z}^+ \text{ such that}$$

$$a - \sum_{j=0}^{n} a_j \in S^{-m} \quad \text{for } n > N$$ i.e. by taking enough of finitely

many terms in the infinite sum, a can be approximated up to an arbitrarily smoothing term.

The next result is crucial.

<u>Theorem</u>. Let Ψ_K denote the class of ΨDO's restricted to functions with support in a fixed compact set K. For $P \in \Psi_K$, let P^* be defined by $(P^*u,v) = (u,P\,v)$ (inner product in L^2). Then

1) $P \in \Psi_K \Rightarrow P^* \in \Psi_K$ and $\sigma(P^*) \sim \sum_\alpha \partial_\xi^\alpha D_x^\alpha p^*/\alpha!$

where $(p(x,\xi))^* = \overline{p(x,\xi)}$.

2) $P,Q \in \Psi_K \Rightarrow PQ \in \Psi_K$ and $\sigma(PQ) = \sum_\alpha \partial_\xi^\alpha p D_x^\alpha q/\alpha!$

We need a

Technical Lemma:

Let $r(x,\xi,y)$ be of compact x and y support and be such that $\forall\, \alpha,\beta,\gamma \quad \exists$ constant $c_{\alpha\beta\gamma}$ with $|D_x^\alpha D_\xi^\beta D_y^\gamma r(x,\xi,y)| \leq c_{\alpha\beta\gamma}(1+|\xi|)^{m-|\beta|}$ for some fixed $m \in \mathbb{R}$.

Then $(Ru)(x) = \int e^{i(x-y)\cdot\xi} r(x,\xi,y)u(y)dyd\xi$ defines an element of Ψ_K and

$$\sigma R \sim \sum_\alpha \partial_\xi^\alpha D_y^\alpha r(x,\xi,y)/\alpha! \Big|_{y=x}.$$

Proof of theorem:

1) $(Pu,v) = \int e^{ix\cdot\xi}\hat{u}(\xi)\bar{v}(x)p(x,\xi)d\xi dx$

$$= \int e^{i(x-y)\cdot\xi}u(y)\bar{v}(x)p(x,\xi)dyd\xi dx.$$

Since $p(x,\xi)$ can be approximated by functions of compact ξ-support, we can apply Fubini's theorem and get

$$(Pu,v) = \int e^{i(x-y)\cdot\xi}u(y)\bar{v}(x)p(x,\xi)dxd\xi dy.$$

Now define $(P^*v)(y) = \int \overline{e^{i(x-y)\cdot\xi}}\, v(x)\overline{p(x,\xi)}dxd\xi$.

Then $(u,P^*v) = (Pu,v)$.

Choose $\phi \in C_0^\infty$ such that $\phi \equiv 1$ on K

$$(P^*v)(x) = \int e^{i(x-y)\cdot\xi}\phi(x)\overline{p(y,\xi)}v(y)dyd\xi ..$$

Let $r(x,\xi,y) = \phi(x)\overline{p(y,\xi)}$.

Then $r(x,\xi,y)$ satisfies the conditions of the lemma so that $P^* \in \Psi_k$ and $\sigma P^* \sim \sum_\alpha \partial_\xi^\alpha D_y^\alpha \phi(x)\overline{p(y,\xi)}/\alpha! \Big|_{y=x}$

$$= \sum_\alpha \partial_\xi^\alpha D_y^\alpha \overline{p(y,\xi)}/\alpha! \Big|_{y=x}$$

from the way ϕ was chosen.

2) $\qquad Q^*u(x) = \int e^{i(x-y)\cdot\xi}\overline{q(y,\xi)}u(y)dyd\xi \qquad$ (from proof of (1))

$$\therefore Qu(x) = ((Q^*)^*u)(x) = \int e^{i(x-y)\cdot\xi}\tilde{q}(y,\xi)u(y)dyd\xi$$

where $\tilde{q}(y,\xi) = \overline{\sigma(Q^*)}$.

We claim that $\widehat{Qu}(\xi) = \int e^{iy\cdot\xi}\tilde{q}(y,\xi)u(y)dy$ (to check this just take $\mathcal{F}^{-1}$ of both sides).

Therefore

$$(PQ)u(x) = P(Qu(x)) = \int e^{ix\cdot\xi}\, p(x,\xi)\widehat{Qu}(\xi)d\xi$$

$$= \int e^{i(x-y)\cdot\xi}p(x,\xi)\tilde{q}(y,\xi)u(y)dyd\xi \ .$$

Now $r(x,\xi,y) = p(x,\xi)\tilde{q}(y,\xi)$ satisfies the conditions of the technical lemma so that $PQ \in \Psi_k$ and

$$\sigma(PQ) \sim \sum_{\alpha} \partial_\xi^\alpha D_y^\alpha r(x,\xi,y)/\alpha!\,|_{y = x}$$

$$= \sum_{\alpha} \partial_\xi^\alpha D_y^\alpha p(x,\xi)\tilde{q}(y,\xi)/\alpha!\,|_{y = x}$$

$$= \sum_{\alpha} \partial_\xi^\alpha [p(x,\xi) D_y^\alpha \tilde{q}(y,\xi)]/\alpha!\,|_{y = x}$$

$$= \sum_{\alpha} \partial_\xi^\alpha [p(x,\xi) D_x^\alpha \tilde{q}(x,\xi)]/\alpha!.$$

Now apply Leibniz's rule which says that

$$\partial_\xi^\alpha(ab) = \sum_{\alpha=\alpha_1+\alpha_2} \left(\partial_\xi^{a_1} a\right)\left(\partial_\xi^{a_2} b\right)\alpha!/(\alpha_1!\alpha_2!)$$

with $a = p(x,\xi)$ and $b = D_x^{\alpha_2} q(x,\xi)$

and get

$$\sigma(PQ) \sim \sum_{\alpha=\alpha_1+\alpha_2} \left(\partial_\xi^{\alpha_1} p(x,\xi)\right)\left(\partial_\xi^{\alpha_2} D_x^{\alpha_1} D_x^{\alpha_2} \tilde{q}(x,\xi)\right)\frac{\alpha!}{\alpha_1!\alpha_2!\alpha!}$$

$$= \sum_{\alpha_1} \partial_\xi^{\alpha_1} p(x,\xi) D_x^{\alpha_1}\left(\sum_{\alpha_2} d_\xi^{\alpha_2} D_x^{\alpha_2} \tilde{q}(x,\xi)/\alpha_2!\right)/\alpha_1!$$

$$= \sum_{\alpha} \partial_\xi^\alpha p(x,\xi) D_x^\alpha q(x,\xi)/\alpha!$$

by (1).

Proof of technical lemma.

$$Ru(x) = \int e^{i(x-y)\cdot\xi} r(x,\xi,y)u(y)dyd\xi .$$

Now $\int e^{-iy\cdot\xi} r(x,\xi,y)u(y)dy = \widehat{r(x,\xi,y)u(y)}(\xi) = (\hat{r}_y(x,\xi,\cdot) * \hat{u}(\cdot))(\xi)$

where $\hat{r}_y$ denotes $\mathcal{F}$ in the y variable.

Therefore $Ru(x) = \int e^{ix\cdot\xi}\hat{r}_y(x,\xi,\xi-\eta)\hat{u}(\eta)d\eta d\xi$.

Now $|\hat{r}_y(x,\xi,\xi-\eta)| \leqslant C_k(1+|\xi|)^m(1+|\xi-\eta|)^{-k}$ and $|\hat{u}(\eta)| \leqslant C_k'(1+|\eta|)^{-k}$ for all k.

So we can apply Fubini to get

$$Ru(x) = \int e^{ix\cdot\xi}\hat{r}_y(x,\xi,\xi-\eta)\hat{u}(\eta)d\xi d\eta .$$

Now define $p(x,\eta) = \int e^{ix\cdot(\xi-\eta)}\hat{r}_y(x,\xi,\xi-\eta)d\xi$

Then $Ru(x) = \int e^{ix\cdot\eta}p(x,\eta)\hat{u}(\eta)d\eta$.

Observe that $p(x,\eta) \in S^m$ because of the properties of $r(x,\xi,y)$, from which it follows that $R \in \Psi_K$. If we change the variable ξ to $\xi+\eta$ then

$$p(x,\eta) = \int e^{ix\cdot\xi}\hat{r}_y(x,\eta+\xi,\xi)d\xi.$$

But $\hat{r}_y$ has a Taylor series expansion in the middle variable

$$\hat{r}_y(x,\eta+\xi,\xi) = \sum_{|\alpha|\leqslant k} \partial_\eta^\alpha\hat{r}_y(x,\eta,\xi)\xi^\alpha/\alpha! + \text{Remainder}$$

Thus $p(x,\eta) = \sum_{|\alpha|\leq k} \int e^{ix\cdot\xi}\partial_\eta^\alpha \hat{r}_y(x,\eta,\xi)\xi^\alpha d\xi/\alpha! + \text{Remainder}$

$= \sum_{|\alpha|\leq k} \partial_\eta^\alpha(\int e^{ix\cdot\xi}\xi^\alpha \hat{r}_y(x,\eta,\xi)d\xi)/\alpha! + \text{Remainder}$

$= \sum_{|\alpha|\leq k} \partial_\eta^\alpha D_y r(x,\eta,y)/\alpha!\Big|_{y=x} + \text{Remainder}$

One can check that the remainder $\in S^{m-k}$. Since $\sigma R = p$, this proves the second part of the lemma. A differential operator P is local in the sense that if $u \equiv 0$ on an open set 0 then $Pu \equiv 0$ on 0 since differentiation is a purely local process. However, ψDO's are not necessarily local as they are defined by integrals, so that the values of a function at distant points can have an effect. Nevertheless ψDO's enjoy a similar but weaker property viz. they are pseudo-local. This means that if P is a ψDO and $u \in C^\infty(0)$ then $Pu \in C^\infty(0)$ for any open set 0. Alternatively, if sing supp u denotes the set of points where u fails to be C^∞, then P is pseudo-local if sing supp $Pu \subseteq$ sing supp u.

Theorem: If P is a ψDO then P is pseudo-local.

Proof: First observe that multiplication by $c(x) \in C_0^\infty$ is a ψDO since $c(x)u(x) = c(x)\int e^{ix\cdot\xi}\hat{u}(\xi)d\xi = \int e^{ix\cdot\xi}c(x)\hat{u}(\xi)d\xi$. Let 0 be an open set and $u \in C^\infty(0)$. Fix $x \in 0$, choose $\phi \in C_0^\infty(0)$ so that $\phi \equiv 1$ near x and $\psi \in C_0^\infty(0)$ so that $\psi \equiv 1$ on supp ϕ.

Now $\phi P\psi u \in C^\infty$ since $\psi u \in C_0^\infty$.

We have $\sigma(\phi P\psi) = \phi\sigma(P\psi) =$

$$\phi \sum_\alpha \partial_\xi^\alpha p D_x^\alpha \psi/\alpha!$$

$= \phi p$ since $\psi \equiv 1$ on supp ϕ

and $D_x^\alpha\psi \equiv 0$ on supp ϕ for $|\alpha| > 0$.

Therefore $\quad \sigma\phi P\psi - \phi P \in S^{-k}$ for all $k > 0$

$\Rightarrow$ $\phi P\psi - \phi\ P: H_s \to C^\infty$ for all s (Sobolev)

$\Rightarrow$ $(\phi P\psi - \phi P)u \in C^\infty$.

$\Rightarrow$ $(\phi P)u \in C^\infty$

$\Rightarrow$ $Pu \in C^\infty$ near x. But x was arbitrary in 0.

A differential operator is a finite sum of operators of decreasing order. The next theorem yields the analogous result that infinite sums of ψDO's of decreasing order define ψDO's.

<u>Theorem</u>: Let $p_j \in S^{m_j}$, $m_j \searrow -\infty$ and supp p_j contained in some fixed compact set for all j. Then there is a unique (up to infinitely smoothing term) $\quad p \sim \sum_{j=0}^\infty p_j$.

<u>Proof</u>: Let $\psi: R^N \to [0,1]$ be a C^∞ function such that $\psi \equiv 0$ on $|\xi| < 1$ and $\psi \equiv 1$ on $|\xi| > 2$ i.e. ψ looks like.

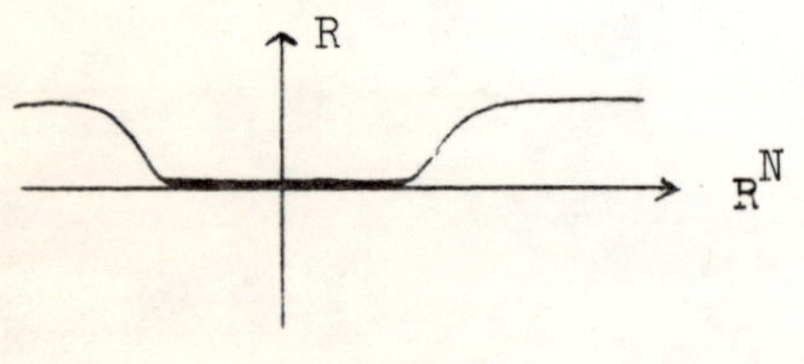

Choose a sequence $t_j \in R^+$ s.t. $t_j \to \infty$.

The sum $\sum_{j=0}^{\infty} \psi(\xi/t_j)p_j(x,\xi)$ is certainly well-defined, since for any fixed (x,ξ) there are only finitely many nonzero terms, $\psi(\xi/t_j)$ eventually being zero.

Suppose $j > 0$. Then

$$\begin{aligned} |p_j(x,\xi)| &\leqslant C(1+|\xi|)^{m_j} \quad \text{since } p_j \in S^{m_j} \\ &= C(1+|\xi|)^{m_j-m_0}(1+|\xi|)^{m_0} \\ &\leqslant \frac{1}{2^j}(1+|\xi|)^{m_0} \quad \text{for } |\xi| \text{ large enough.} \end{aligned}$$

By passing to a subsequence of the t_j's we can assume that

$$|\psi(\xi/t_j)p_j(x,\xi)| \leqslant \frac{1}{2^j}(1+|\xi|)^{m_0} \quad \text{for \underline{all} } \xi .$$

Therefore

$$\begin{aligned} \Big|\sum_{j=0}^{\infty} \psi(\xi/t_j)p_j(x,\xi)\Big| &\leqslant |p_0(x,\xi)| + (1+|\xi|)^{m_0} \\ &\leqslant C(1+|\xi|)^{m_0} \quad \text{for suitable } C. \end{aligned}$$

Using essentially the same method we can show that in fact

$$\sum_{j=0}^{\infty} \psi(\xi/t_j)p_j(x,\xi) \in S^{m_0} .$$

For $j > 1$ one shows similarly that $|\psi(\xi/t_j)p_j(x,\xi)| \leq \frac{1}{2^j}(1+|\xi|)^{m_1}$, passing to a subsequence of the t_j's if necessary, so that

$$\sum_{j=1}^{\infty} \psi(\xi/t_j)p_j(x,\xi) \in S^{m_1} .$$

By a diagonalization type argument one gets for a suitable subsequence of the t_j's that

$$\sum_{j=k}^{\infty} \psi(\xi/t_j)p_j(x,\xi) \in S^{m_k} \quad \text{for all } k.$$

After this diagonalization process define

$$p(x,\xi) = \sum_{j=0}^{\infty} \psi(\xi/t_j)p_j(x,\xi).$$

Observe that for each j, $\psi(\xi/t_j)p_j(x,\xi)-p_j(x,\xi)$ has compact ξ -support and therefore lives in S^k for all k.

Let $\phi = \psi - 1$. Then $\phi(\xi/t_j)p_j(x,\xi)$ is infinitely smoothing for all j.

Now $p(x,\xi) \in S^{m_0}$

$$\begin{aligned} p(x,\xi)-p_0(x,\xi) &= \phi(\xi/t_j)p_0(x,\xi) + \sum_{j=1}^{\infty} \psi(\xi/t_j)p_j(x,\xi) \\ &\sim \sum_{j=1}^{\infty} \psi(\xi/t_j)p_j(x,\xi) \in S^{m_1} \end{aligned}$$

$$p(x,\xi)-p_0(x,\xi)-\dots-p_k(x,\xi) = \sum_{j=0}^{k} \phi(\xi/t_j)p_j(x,\xi) + \sum_{j=k+1}^{\infty} \psi(\xi/t_j)p_j(x,\xi)$$

$$\sim \sum_{j=k+1}^{\infty} \psi(\xi/t_j)p_j(x,\xi) \in S^{m_{k+1}}$$

which proves that $p \sim \sum_{j=0}^{\infty} p_j$.

Elliptic operators.

So far we have constructed the algebra of ψDO's closed under composition and adjoint. It turns out that the "invertible" elements in this algebra are extremely useful tools--these are called elliptic. More precisely:

Definition: A symbol $p(x,\xi) \in S^m$ is said to be elliptic if $p(x,\xi)^{-1}$ exists and there is a constant C such that $|p(x,\xi)^{-1}| \leq C(1+|\xi|)^{-m}$, for large ξ. A ψDO P is said to be elliptic if its symbol p is elliptic.

Examples:

(1) The Laplacian Δ on functions in R^N

$$\Delta f = \sum_{i=1}^{N} \frac{\partial^2 f}{\partial x_i^2} \quad , \quad \sigma(\Delta) = -|\xi|^2$$

(2) The operator $\frac{\partial}{\partial \bar{z}}$ on functions in C

$$\frac{\partial f}{\partial z} = \frac{1}{2}\left(\frac{\partial f}{\partial x} + i \frac{\partial f}{\partial y}\right) \quad , \sigma\left(\frac{\partial}{\partial \bar{z}}\right) = \frac{1}{2i} (\xi + i\eta) .$$

If $P = \sum_{|\alpha| \leq m} a_\alpha(x) D^\alpha$ is an elliptic differential operator, the leading order symbol $\sigma_L P = \sum_{|\alpha|=m} a_\alpha \xi^\alpha$. There is a nice simple criterion for ellipticity in the case of differential operators.

<u>Theorem</u>: Let $P = \sum_{|\alpha|\leq m} a_\alpha(x)D^\alpha$.

Then P is elliptic $\Leftrightarrow \sigma_L P(x,\xi) \neq 0$ for all $\xi \neq 0$.

<u>Proof</u>: Let P be elliptic, then for some $\varepsilon > 0$,

$$\varepsilon|\xi|^m \leq \Big|\sum_{|\alpha|\leq m} a_\alpha(x)\xi^\alpha\Big| \quad \text{for } |\xi| \text{ large.}$$

However, $\Big|\sum_{|\alpha|\ m} a_\alpha(x)\xi^\alpha\Big| \leq |\sigma_L(P)(x,\xi)| + \Big|\sum_{|\alpha|\leq m-1} a_\alpha(x)\xi^\alpha\Big|$

$$\leq |\sigma_L(P)(x,\xi)| + \varepsilon/2\ |\xi|^m \quad \text{if } \xi \text{ is large.}$$

Therefore: $\frac{\varepsilon}{2}|\xi|^m \leq |\sigma_L(P)(x,\xi)|$ for ξ-large. However, $\sigma_L P$ is homogeneous of order m and therefore $\sigma_L(P)(x,\eta) \neq 0$ for all $\xi \neq 0$.

Conversely, suppose that $\sigma_L P(x,\xi) \neq 0$ for all $\xi \neq 0$. Now $|\sigma_L P(x,\xi)| \geq c|\xi|^m$ for some constant c. But $p(x,\xi) = \sum_{|\alpha|=m} a_\alpha(x)\xi^\alpha + \sum_{|\alpha|<m} a_\alpha(x)\xi^\alpha$.

Therefore

$$\left|\frac{p(x,\xi)}{\sigma_L P(x,\xi)} - 1\right| \leq \sum_{|\alpha|<m}\left|\frac{a_\alpha(x)\xi^\alpha}{\sigma_L P(x,\xi)}\right| \leq c \sum_{|\alpha|<m} \frac{|a_\alpha(x)\xi^\alpha|}{|\xi|^m} \to 0 \quad \text{as } |\xi| \to \infty$$

so that $p(x,\xi)$ is invertible for large ξ. Given $\varepsilon > 0$, for ξ large we have $\sum_{|\alpha|<m} |a_\alpha(x)\xi^a| < \varepsilon|\xi|^m$ so that

$$|p(x,\xi)^{-1}| = \left|\frac{1}{\sigma_L P(x,\xi) + \sum_{|\alpha|<m} a_\alpha(x)\xi^\alpha}\right| \leq \frac{1}{c|\xi|^m - \varepsilon|\xi|^m} = \frac{1}{c-\varepsilon}\,|\xi|^{-m}$$

$$\Rightarrow \quad |p(x,\xi)^{-1}| \leqslant c(1+|\xi|)^{-m} \text{ for large } \xi.$$

<u>Corollary</u>: Let $p(x,\xi) = p_0(x,\xi) + p_1(x,\xi)$ where $p_0 \in S^m$ and $p_1 \in S^{m-1}$ are symbols of differential operators. Then p is elliptic $\Leftrightarrow p_0$ is elliptic.

The next step is to construct the inverse ψDO of an elliptic ψDO.

<u>Lemma</u>: If $p(x,\xi) \in S^m$ is elliptic and $p(x,\xi)^{-1}$ exists for $|\xi| \geqslant c$ then $p(x,\xi)^{-1}$ can be extended in a C^∞ way to $q(x,\xi) \in S^{-m}$ so that $pq \sim qp \sim 1$.

<u>Proof</u>: Choose a C^∞ function $\psi: R^N \to [0,1]$ such that $\psi \equiv 0$ for $|\xi| < c + 1$ and $\psi \equiv 1$ for $|\xi| > c + 2$.

Define

$$q(x,\xi) = \begin{cases} 0 & \text{for } |\xi| < c + 1 \\ \psi(\xi)p(x,\xi)^{-1} & \text{for } |\xi| \geqslant c + 1. \end{cases}$$

Then $pq - 1 \in C_0^\infty$ and $qp - 1 \in C_0^\infty$ so that $pq \sim qp \sim 1$. It remains to show that $q \in S^{-m}$.

Clearly $|q(x,\xi)| \leqslant c(1+|\xi|)^{-m}$

$$\partial_\xi^\alpha q(x,\xi) = 0 \quad \text{when } |\xi| < c + 1$$

$$\partial_\xi^1(\psi/p) = \frac{\partial_\xi^1\psi}{p} - \frac{\psi\partial_\xi^1 p}{p^2} \quad \text{when } |\xi| \geqslant c + 1.$$

We can ignore the effect of $\dfrac{\partial_\xi^1 \psi}{p}$ for $|\xi| \geq c + 1$

and $\psi \partial_\xi^1 p \in S^{-m-1}$
$p^2 \in S^{2m}$ $\Rightarrow \dfrac{\psi \partial_\xi^1 p}{p^2} \in S^{-m-1} \Rightarrow \partial_\xi^1 q \in S^{-m-1}$.

Similarly one shows that $D_x^1 \partial_\xi^1 q(x,\xi) \in S^{-m-1}$, i.e. differentiating in the x-direction leaves the order unchanged. Computing higher derivatives yields

$$|D_x^\alpha \partial_\xi^\beta q(x,\xi)| \leq c_{\alpha\beta}(1+|\xi|)^{-m-|\beta|}.$$

<u>Theorem</u>: If P is an elliptic operator then there is a unique (up to an infinitely smoothing operator) ψDO Q such that $PQ \sim I \sim QP - I \sim 0$.

<u>Proof</u>: By the lemma we can assume that $p(x,\xi)^{-1}$ exists for all ξ. We work backwards and assume that we have found Q such that $\sigma Q = q = q_0 + q_1 + \ldots$ with $q_i \in S^{-m-i}$.

Then

$$\sigma(PQ) \sim \sum_{\alpha,j} \partial_\xi^\alpha p D_x q_j/\alpha! \text{ where } \partial_\xi^\alpha p \in S^{m-|\alpha|}, D_x^\alpha q_j \in S^{-m-i}$$

$$= \sum_{k=0}^\infty \sum_{|\alpha|+j=k} \partial_\xi^\alpha p D_x^\alpha q_j/\alpha! \text{ where } \partial_\xi^\alpha p D_x^\alpha q_j \in S^{-k} .$$

For $k = 0$ there is only one term $,pq_0$. Set $q_0 = \frac{1}{p} \in S^{-m}$. For $k > 0$ we want $\sum_{|\alpha|+j=k} \partial_\xi^\alpha p D_x^\alpha q_j/\alpha! = 0$.

Assume that q_j has been defined for $j < k$.

Now $\sum_{|\alpha|+j=k} \partial_\xi^\alpha p D_x^\alpha q_j/\alpha! = pq_k + \sum_{\substack{|\alpha|+j=k \\ j<k}} \partial_\xi^\alpha p D_x^\alpha q_j/\alpha!$. Inductively define

$$q_k = -\frac{1}{p} \sum_{\substack{|\alpha|+j=k \\ j<k}} \partial_\xi^\alpha p D_x^\alpha q_j/\alpha! \in S^{-m-k} .$$

Let $q \sim \sum_{k=0}^{\infty} q_k$ (q exists by a previous theorem).

By construction we then have

$$\sigma(PQ) \sim pq + \text{lower order terms} = 1.$$

Therefore $PQ - I \sim 0$. Similarly one can construct $\tilde{Q}$ such that $\tilde{Q}P - I \sim 0$. But $\tilde{Q} \sim \tilde{Q}(PQ) \sim (\tilde{Q}P)Q \sim Q$, so that Q is unique as claimed.

The previous result has an important corollary which yields many results about regularity.

<u>Corollary (1)</u>. If P is elliptic and $Pu = 0$ then $u \in C^\infty$. In particular harmonic and analytic functions are C^∞.

<u>Proof</u>: $Pu = 0$. Now $(QP - I)u \in C^\infty$ so

$$QPu - u = -u \in C^\infty \Rightarrow u \in C^\infty$$

(2) If P is elliptic and $Pu \in C^\infty$ on an open set V then $u \in C^\infty(V)$.

<u>Proof</u>: $(QP - I)u \in C^\infty$ and $Pu \in C^\infty$. Now $QPu \in C^\infty$ (ψDO's are pseudo-local) $\Rightarrow u \in C^\infty(V)$.

Invariance of ψDO's under change of coordinates.

We are going to transfer all the machinery of ψDO's to manifolds. Before we can do that we must see what happens to symbols under change of coordinates in R^N. In some sense we want symbols to be invariant under such changes. Let V be an open subset of R^N and $f:\tilde{V} \to V$ a diffeomorphism defining new coordinates $\tilde{x}$ in terms of old ones x.

For $\phi \in C_0^\infty(V)$ define $\tilde{\phi}(\tilde{x}) = \phi(f(\tilde{x})) = \phi(x)$. For a ψDO P acting on $C_0^\infty(V)$, define $\tilde{P}\tilde{\phi}(\tilde{x}) = (P\phi)f(\tilde{x}) = (P\phi)(x)$.

Example: Let $f(\tilde{x}) = a\tilde{x} + b$, $\det a \neq 0$.

Then $\tilde{P}\tilde{\phi}(\tilde{x}) = (P\phi)(x) = \int e^{i(x-y)\cdot\xi}p(x,\xi)\phi(y)dyd\xi$

$$= \int e^{ia(\tilde{x}-\tilde{y})\cdot\xi}p(a\tilde{x}+b,\xi)\tilde{\phi}(\tilde{y})a\, d\tilde{y}d\xi .$$

Change variables, $a^t\xi \to \eta$ and get

$$\tilde{P}\tilde{\phi}(\tilde{x}) = \int e^{i(\tilde{x}-\tilde{y})\cdot\eta}p(a\tilde{x} + b,(a\ )^{-1}\eta)\tilde{\phi}(\tilde{y})d\tilde{y}d\eta.$$

Thus $\tilde{P}$ is a ϕDO with $\sigma\tilde{P} = \tilde{p}(\tilde{x},\eta) = p(a\tilde{x} + b,(a\ )^{-1}\eta)$.

Notice that under the change of coordinates the ξ variable in $p(x,\xi)$ transforms to $(a\ )^{-1}\eta$. But this is how coordinates for cotangent vectors change under change of coordinates, so in this case P is invariantly defined on $T^*(R^N)$.

Now consider a general change of coordinates given by $x = f(\tilde{x})$

$$f(\tilde{x}) - f(\tilde{y}) = \int_0^1 \frac{d}{dt} f(t\tilde{x} + (1-t)\tilde{x})dt$$

$$= \int_0^1 \frac{\partial}{\partial x} f(t\tilde{x} + (1-t)\tilde{y})(\tilde{x}-\tilde{y})dt \qquad \text{(chain rule)}$$

$$= H(\tilde{x},\tilde{y})(\tilde{x}-\tilde{y}).$$

Observe that $H(\tilde{x},\tilde{x}) = \frac{\partial f}{\partial x}$ which must be a nonsingular matrix since f defines a change of coordinates. By continuity of determinant, $\det H(\tilde{x},\tilde{y}) \neq 0$ for $\tilde{x}$, $\tilde{y}$ sufficiently close. Since we are going to be using partitions of unity anyway on manifolds, we suffer no loss of generality if we assume that p has very small support so that all $\tilde{x}$ and $\tilde{y}$ will be close enough for $\det H(\tilde{x},\tilde{y}) \neq 0$.

$$\tilde{P}\tilde{\phi}(\tilde{x}) = P\phi(x) = \int e^{i(x-y)\cdot\xi} p(x,\xi)\phi(y)dyd\xi$$

$$= \int e^{i(f(\tilde{x})-f(\tilde{y}))\cdot\xi} p(f(\tilde{x}),\xi)\tilde{\phi}(\tilde{y})dyd\xi$$

$$= \int e^{i(\tilde{x}-\tilde{y})\cdot H^t(\tilde{x},\tilde{y})\cdot\xi} p(f(\tilde{x}),\xi)\tilde{\phi}(\tilde{y})\left|\det\frac{\partial f}{\partial y}\right| d\tilde{y}d\xi.$$

Change variables by setting $H^t(\tilde{x},\tilde{y})\cdot\xi = \eta$ and get

$$\tilde{P}\tilde{\phi}(\tilde{x}) = \int e^{i(\tilde{x}-\tilde{y})\cdot\eta} p(f(\tilde{x}),(H^t(\tilde{x},\tilde{y}))^{-1}\eta)\tilde{\phi}(\tilde{y})\left|\det\frac{\partial t}{\partial y}\right|\left|\det H^t(\tilde{x},\tilde{y})^{-1}\right| d\tilde{y}d\eta.$$

Let $r(\tilde{x},\eta,\tilde{y}) = p(f(\tilde{x}),(H^t(\tilde{x},\tilde{y}))^{-1}\eta)|\det H^t(\tilde{x},\tilde{y}))^{-1}|\ |\det \frac{\partial f}{\partial y}|$.

Then r satisfies the conditions of the technical lemma, so that $\tilde{P}$ defines a ψDO and

$$\sigma\tilde{P} \sim \sum_{|\alpha|\geq 0} \partial_\xi^\alpha D_y^\alpha r(\tilde{x},\eta,\tilde{y})/\alpha!\ |_{\tilde{y}=x}.$$

Define the <u>leading order symbol</u> of a ψDO $\sum_{j=0}^{\infty} a_j$ with $a_j \in S^{m_j}$, $m_j \searrow -\infty$ to be a_0.

Thus the leading order symbol of $\sigma\tilde{P}$ is the term obtained for $\alpha = 0$, i.e. $r(\tilde{x},\eta,\tilde{x}) = p(f(\tilde{x}),(H^t(\tilde{x},\tilde{x}))^{-1}\eta) = p(f(\tilde{x}), \left(\left(\frac{\partial f}{\partial \tilde{x}}\right)^t\right)^{-1}\eta)$.

We have proved the following:

<u>Theorem</u> (Kuranishi): The class of ψDO's with very small x-support have their leading order symbols invariantly defined on $T^*(R^N)$.

CHAPTER II: Elliptic Operators on Compact Manifolds

From now on M will be a C^∞ compact Riemannian manifold without boundary, dimension $(M) = N$. The volume element of M will be denoted by dvol.

Definition. Let $C^\infty(M)$ be the smooth complex-valued functions on M. $P: C^\infty(M) \to C^\infty(M)$ is said to be a pseudo-differential operator on $C^\infty(M)$ if the induced map $P_{\phi_\alpha} : C^\infty(\phi_\alpha(U_\alpha)) \to C^\infty(\phi_\alpha(U_\alpha))$ is a ψDO for each coordinate chart $\phi_\alpha : U_\alpha \to R^N$ of M. ψDO's on compact manifolds certainly exist -- simply patch together locally Euclidean ψDO's with a partition of unity.

Proposition. For each $m \in \mathbb{R}$, there is an elliptic operator of order m on M.

Proof: Let $\{U_\alpha\}$ be an atlas for M.

Define D_α on U_α be setting $\sigma D_\alpha = (1+|\xi|_\alpha)^m$ where $\xi_\alpha = (\xi_1,\ldots,\xi_N)$ are coordinates in T^*U_α. Let $\{\rho_\alpha\}$ be a partition of unity on M which is locally finite and subordinate to $\{U_\alpha\}$. If we define D on all of M by setting $D = \sum_\alpha \rho_\alpha D_\alpha$, then D will be elliptic of order m.

We would like to define H_s spaces on the manifold. The obvious way to do this is to let

$$\|u\|_s = \left\{ \sum_\alpha \|\rho_\alpha u\|_s^2 \right\}^{1/2} \text{ where } \{\rho_\alpha\} \text{ is a partition of unity}$$

sub-ordinate to a finite covering $\{U_\alpha\}$ of M by charts; $u \in C^\infty(M)$ Let $H_s(M)$ be the completion of $C^\infty(M)$ in this norm. (One should think of $\|\rho_\alpha u\|_s$ as the norm of the induced function on $\phi_\alpha(U_\alpha) \subset R^N$). However, this method of defining $H_s(M)$ is not canonical except in the case $s = 0$ when we can simply define

$$L^2(M) = \{u: \int_M |u|^2 dvol < \infty\}.$$

We shall show shortly that in fact the H_s spaces are independent of any choices made, up to equivalent norms.

First we observe that all the theorems we proved about ψDO's in $\mathbb{R}^N$ are true also for ψDO s in M. This is true because ψDO's in M are defined to be locally ψDO's in $\mathbb{R}^N$. So we have

(1) $H_s(M) \hookrightarrow H_t(M)$ is a compact map for $s > t$
(2) $\phi \in H_{s+k}(M) \Rightarrow \phi \in C^k(M)$ for $s > N/2$
(3) $\sigma P \in S^m \Rightarrow P: H_s(M) \to H_{s-m}(M)$ is continuous
(4) $H_s(M) \times H_{-s}(M) \to \mathbb{C}$ is a perfect pairing
(5) $\sigma_L P$ is invariantly defined on T^*M.

The following proposition establishes that $H_s(M)$ is defined independently of the choices made, up to equivalent norms.

<u>Proposition</u>. If $s \geqslant 0$ an equivalent norm for $H_s(M)$ is $\|u\|_s = \|u\|_0 + \|Pu\|_0$ where P is any elliptic operator of order s.

Proof: Given $s \geqslant 0$, by the previous proposition there is an elliptic operator in M, say P, of order s. Let Q be a ψDO in M of order $-s$ such that $QP - I \sim 0$. Then

$$\|u\|_s \leqslant \|QPu\|_s + \|(QP - I)u\|_s$$

$$\leqslant \|QPu\|_s + C\|u\|_0 \,, \quad \text{since } QP - I \sim 0$$

$$\leqslant C\|Pu\|_0 + C\|u\|_0 \,, \quad \text{since } Q \text{ is of order } -s$$

$$= C(\|u\|_0 + \|Pu\|_0).$$

On the other hand

$$\|Pu\|_0 \leqslant C\|u\|_s \,, \quad \text{since } P \text{ is of order } s$$

$$\therefore \|Pu\|_0 + \|u\|_0 \leqslant C\|u\|_s \,.$$

Now define $H_{-s} = (H_s)^*$ for $s \geqslant 0$, and the proposition extends to the negative case.

We can generalize the previous results about ψDO's on manifolds to vector bundles over M.

Let E be a complex vector bundle over M provided with an Hermitian inner product. $C^\infty(E)$ denotes the smooth sections of E.

<u>Definition.</u> $P:C^\infty(E) \to C^\infty(F)$ is a <u>ψDO of order m</u> if locally P is a matrix such that each component of the matrix is a ψDO of order m.

In this case $\sigma P: T_x^*M \to \mathrm{Hom}(E_x,F_x)$ is a matrix for each $\xi \in T_x^*M$, where E_x,F_x are the fibers of E and F over x.

P is elliptic $\Leftrightarrow$ $\sigma_L P$ maps E_x isomorphically onto F_x for each $x \in X$ and each $\xi \in T_x^*M$, $\xi \neq 0$.

Let $(\, , \,)_x$ be the Hermitian inner product on E_x and define $\langle\phi,\psi\rangle = \int_M (\phi,\psi)_x d\mathrm{vol}$ for $\phi, \psi \in C^\infty(E)$. $\langle \, , \, \rangle$ induces a norm and $L^2(E)$ is defined to be the completion of $C^\infty(E)$ in this norm.

$H_s(E)$ can be defined similarly to the way in which it was defined for $C^\infty(M)$, using partitions of unity. The only complication is that locally we have to deal with vector-valued functions rather than scalar valued functions.

Having made these definitions, it is not hard to show that all the results we obtained earlier about ψDO's are true also for vector bundles over M

e.g. $H_s(E) \hookrightarrow H_t(E)$ is a compact map for $t > s$ if P is of order m, $P:H_s(E) \to H_{s-m}(F)$ is continuous, etc.

Elliptic Complexes.

The elliptic operators we shall be studying will arise from elliptic complexes over manifolds.

An elliptic complex $(\mathcal{E}, d)$ over M is a finite collection $\{E_i\}_{i=0}^n$ of vector bundles over M with a differential operator $d_i : C^\infty(E_i) \to C^\infty(E_{i+1})$ of order $m > 0$ for each $i = 0, \ldots, n-1$ satisfying the conditions

(1) $d_{i+1} d_i = 0$

(2) For each $x \in M$ and each $\xi \in T_x^*(M)$, $\xi \neq 0$, the sequence

$$0 \to E_0 \xrightarrow{\sigma_L d_0(x,\xi)} E_1 \xrightarrow{\sigma_L d_1(x,\xi)} E_2 \to \cdots \xrightarrow{\sigma_L d_{n-1}(x,\xi)} E_n \to 0$$

is exact.

Notice that a two-step elliptic complex $d_0 : C^\infty(E_0) \to C^\infty(E_1)$ is simply an elliptic operator.

Given an elliptic complex $\mathcal{E}$ we construct elliptic operators

$$D_i : C^\infty(E_i) \to C^\infty(E_i)$$

by setting

$$D_i = d_{i-1} d_{i-1}^* + d_i^* d_i ,$$

which is clearly self-adjoint. To check that D_i is in fact elliptic for each i we need to show that

$$\sigma_L D_i = \sigma_L d_{i-1} \sigma_L d_{i-1}^* + \sigma_L d_i^* \sigma_L d_i \quad \text{is invertible.}$$

We omit the subscripts in the proof.

Suppose $(\sigma_L D)\theta = 0$ for $\theta \in E$. Then

$$(\sigma_L d \sigma_L d^* + \sigma_L d^* \sigma_L d)\theta = 0$$

$$\Rightarrow ((\sigma_L d \sigma_L d^* + \sigma_L d^* \sigma_L d)\theta, \theta) = 0$$

$$\Rightarrow (\sigma_L d^* \theta, \sigma_L d^* \theta) + (\sigma_L d\theta, \sigma_L d\theta) = 0$$

$$\Rightarrow \sigma_L d^* \theta = 0 = \sigma_L d\theta .$$

Since $\sigma_L(d)\theta = 0$, the exactness shows there is a $\psi \in E_{i-1}$ such that $\theta = (\sigma_L d)\psi$. Thus $0 = (\sigma_L d^*)\theta = (\sigma_L d^*)(\sigma_L d)\psi$. Therefore

$$((\sigma_L d^*)(\sigma_L d)\psi, \psi) = 0$$

$$\Rightarrow (\sigma_L d\psi, \sigma_L d\psi) = 0$$

$$\Rightarrow \sigma_L d\psi = 0$$

$$\Rightarrow \theta = 0$$

$$\Rightarrow \sigma_L D \text{ is invertible.}$$

Given an elliptic complex $(\mathcal{E},d)$, define the <u>cohomology vector spaces</u> $H^i(M,\mathcal{E}) = \ker d_i / \operatorname{im} d_{i-1}$. We shall show shortly that $N(D_i)$ has finite dimension for all i, and we can therefore define

$$\operatorname{index}(d) = \sum_{i=0}^{n} (-1)^i \dim N(D_i).$$

The simplest classical example of an elliptic complex is the <u>DeRham complex</u>

$$0 \to \Lambda^0 T^*M \xrightarrow{d} \cdots \to \Lambda^p T^*M \xrightarrow{d} \Lambda^{p+1} T^*M \xrightarrow{d} \cdots \xrightarrow{d} \Lambda^N T^*M \to 0$$

d is of order one and $d^2 = 0$.

To show that the DeRham complex satisfies the axioms of an elliptic complex, we first compute:

$$\sigma d(x,\xi)(\omega_x) = \sigma_L d(x,\xi)(\omega_x) = i\xi \wedge \omega_x$$

where $\xi \in T_x^*M$ and $\omega_x \in \Lambda^p T_x^*M$.

When $p = 0$, for $f \in \Lambda^0$, $df = \sum_{i=1}^{N} \frac{\partial f}{\partial x_i} dx_i = \sum_{i=1}^{N} \frac{\partial}{\partial x_i} dx_i\, f$.

Therefore $(\sigma d(x,\xi))f = i \sum_{i=1}^{N} \xi_i dx_i \cdot f = i\xi \wedge f$

In general for $\omega \in \Lambda^p$, $\omega = \sum_{i_1 < \cdots < i_p} a_{i_1 \cdots i_p} dx_{i_1} \wedge \cdots \wedge dx_{i_p}$

$$d\omega = \sum da_{i_1\ldots i_p} dx_{i_1} \wedge \ldots \wedge dx_{i_p}$$

$$= \sum_{i_1<\ldots<i_p} \left(\left(\sum_{i=1}^{N} \frac{\partial}{\partial x_i} \right) a_{i_1\ldots i_p} dx_i \wedge dx_{i_1} \wedge \ldots \wedge dx_{i_p} \right) .$$

Therefore $(\sigma d(x,\xi))\omega = i \sum_{i_1<\ldots<i_q} \left(\left(\sum_{i=1}^{N} \xi_i dx_i \right) \cdot a_{i_1\ldots i_p} \wedge dx_{i_1} \wedge \ldots \wedge dx_{i_p} \right)$

$$= i \left(\sum_{i=1}^{N} \xi_i dx_i \right) \wedge \sum_{i_1<\ldots<i_p} a_{i_1\ldots i_p} dx_{i_1} \wedge \ldots \wedge dx_{i_p}$$

$$= i\xi \wedge \omega .$$

Next we must show that the symbol sequence is exact. This is best illustrated by an example.

Suppose e.g. that $\xi = \xi_1 dx_1 + \xi_2 dx_2$ i.e. $\xi_3 = \xi_4 = \ldots = \xi_N = 0$ and $\sigma d(x,\xi)\omega = i\xi \wedge \omega = 0$ for $\omega \in \Lambda^p$. This must mean that ω is of the form

$$\omega = \sum_{1<2<i_3<\ldots<i_p} a_{12i_3\ldots i_p} dx_1 \wedge dx_2 \wedge dx_{i_3} \wedge \ldots \wedge dx_{i_p} .$$

Let $\eta \in \Lambda^{p-1}$ be defined by

$$\eta = -i \cdot \xi_1^{-1} \left(\sum_{2<i_3\ldots<i_p} a_{12i_3\ldots i_p} dx_2 \wedge dx_{i_3} \wedge \ldots \wedge dx_{i_p} \right)$$

Then $\omega = i\xi \wedge \eta = \sigma d(x,\xi)\eta$.

Compact and Fredholm operators.

Let $C: H_1 \to H_2$ be a bounded linear operator from one Hilbert space to another.

C is said to be compact if C maps bounded sets to precompact sets; equivalently, if $\{h_n\} \subseteq H_1$ is such that $\|h_n\| \leq 1$ for all n, then $\{h_n\}$ contains a subsequence $\{h_{n_k}\}$ such that $\{Ch_{n_k}\}$ is Cauchy.

A bounded linear operator $F: H_1 \to H_2$ is said to be Fredholm if there is a $G: H_2 \to H_1$ such that $GF - I$ and $FG - I$ are compact operators. We shall be using the following well-known result about Fredholm operators.

Theorem:

$F: H_1 \to H_2$ is Fredholm

$$\Longleftrightarrow \left\{\begin{array}{l} \dim N(F) < \infty \\ \dim N(F^*) < \infty \\ R(F) \text{ is closed.} \end{array}\right\} \Leftrightarrow \left\{\begin{array}{l} H_1 = N(F) \oplus R(F^*) \\ H_2 = N(F^*) \oplus R(F) \end{array}\right\}$$

For a proof see e.g. Lang: Analysis II.

The Hodge Decomposition Theorem:

We are now in a position to give an easy proof of Hodge's decomposition theorem.

Theorem: Let E be a smooth vector bundle over M with Hermitian inner product and

$$D: C^{\infty}(E) \to C^{\infty}(E) \quad \text{a self-adjoint elliptic operator of order } m > 0.$$

Then $L^2(E) = N(D) \oplus R(D)$ with $N(D) \subseteq C^{\infty}(E)$, $\dim N(D) < \infty$ and $C^{\infty}(E) = N(D) \oplus D(C^{\infty}(E))$. (Orthogonal direct sum.)

Proof. D is of order $m \Rightarrow D: H_s(E) \to H_{s-m}(E)$ for all s

D is elliptic $\Rightarrow$ there is a $B: H_{s-m}(E) \to H_s(E)$ such that

$BD - I \sim DB - I \sim 0 \Rightarrow BD - I: H_s \to H_t$ for all s and t.

But $H_t \hookrightarrow H_s$ is a compact injection for $t > s$. Therefore the map $H_s \xrightarrow{BD-I} H_t \hookrightarrow H_s$ is compact, $t > s$, i.e. $BD - I: H_s(E) \to H_s(E)$ is compact. Similarly $DB - I: H_s(E) \to H_s(E)$ is compact. Therefore, $D: H_s(E) \to H_{s-m}(E)$ is Fredholm. In particular $D: H_m(E) \to H_0(E) = L^2(E)$ is Fredholm. Therefore $L^2(E) = N(D^*) \oplus R(D)$. $N(D^*) = N(D)$ since D is self-adjoint. Since D is Fredholm, $N(D)$ is finite. Therefore:

$$L^2(E) = N(D) \oplus R(D) \qquad R(D) = \text{range } (D).$$

By one of the regularity theorems we proved in Chapter I $D\phi = 0 \Rightarrow \phi \in C^{\infty}(E)$ so that $N(D) \subseteq C^{\infty}(E)$.

Now suppose $\phi \in C^{\infty}(E) \subseteq L^2(E)$. By what we have just proved $\phi = \phi_1 + D\phi_2$ where $\phi_1 \in N(D) \subseteq C^{\infty}(E)$ so that $D\phi_2 \in C^{\infty}(E) \Rightarrow \phi_2 \in C^{\infty}(E)$ by one of the regularity theorems. Therefore $C^{\infty}(E) = N(D) \oplus D(C^{\infty}(E))$.

We can apply the previous result to the elliptic operators associated with an elliptic complex.

Theorem: Let $(\mathcal{E},d)$ be an elliptic complex, $D_i = d_{i-1}d^*_{i-1} + d^*_i d_i$. Then $L^2(E_i) = N(D_i) \oplus R(d_{i-1}) \oplus R(d^*_i)$ and there is a 1-1 correspondence between $N(D_i)$ and $H^i(M,\mathcal{E})$.

Proof. By the previous theorem every element $\phi \in L^2(E_i)$ is of the form $\phi = \tilde{\phi} + (d_{i-1}d^*_{i-1} + d^*_i d_i)\psi$ with $\tilde{\phi} \in N(D_i)$ and $\psi \in C^\infty(E_i)$. But $(dd^*\psi, d^*d\psi) = (d^2d^*\psi, d\psi) = 0$ since $d^2 = 0$. Therefore

$$L^2(E_i) = N(D_i) \oplus R(d_{i-1}d^*_{i-1}) \oplus R(d^*_i d_i)$$

$$= N(D_i) \oplus R(d_{i-1}) \oplus R(d^*_i) .$$

Now observe that $\phi \in N(D_i) \Leftrightarrow d^*_{i-1}\phi = 0$ and $d_i\phi = 0$ since

$$D_i\phi = 0 \Longleftrightarrow (d_{i-1}d^*_{i-1} + d^*_i d_i)\phi = 0$$

$$\Longleftrightarrow ((d_{i-1}d^*_{i-1} + d^*_i d_i)\phi, \phi) = 0$$

$$\Longleftrightarrow (d^*_{i-1}\phi, d^*_{i-1}\phi) + (d_i\phi, d_i\phi) = 0$$

Thus we can define the map $N(D_i) \overset{\mu}{\to} H^i(M,\mathcal{E})$

$$\phi \mapsto [\phi]$$

since $\phi \in N(D_i) \Rightarrow d_i\phi = 0$. 1) μ is onto: Suppose ψ ker d_i, $\psi \in C^\infty(E_i)$. Then $\psi = \tilde{\psi} + d_{i-1}\psi_1 + d^*_i\psi_2$ where $\tilde{\psi} \in N(D_i)$ and $0 = d_i\psi = d_i\tilde{\psi} + d_i d_{i-1}\psi_1 + d_i d^*_i\psi_2 = d_i d^*_i\psi_2 \Rightarrow d^*_i\psi_2 = 0 \Rightarrow \psi = \tilde{\psi} + d_{i-1}\psi_1$, so let $\mu(\tilde{\psi}) = [\tilde{\psi}] = [\psi]$. 2) μ is 1-1: Suppose $[\phi_1] = [\phi_2]$ where $\phi_1, \phi_2 \in N(D_i)$ then $\phi_1 = \phi_2 + d_i\psi_1$. Therefore $0 = d^*_{i-1}\phi_1 = d^*_{i-1}\phi_2 + d^*_{i-1}d_{i-1}\psi = d^*_{i-1}d_{i-1}\psi \Rightarrow d_{i-1}\psi = 0 \Rightarrow \phi_1 = \phi_2$. Thus we have exhibited a 1-1 correspondence between $N(D_i)$ and $H^i(M,\mathcal{E})$.

Corollary: $\dim H^i(M,\mathcal{E}) < \infty$. If we take $(\mathcal{E},d)$ to be the DeRham complex we get the classical Hodge Decomposition Theorem:

(a) $L^2(\Lambda^p) = \{\text{harmonic p-forms}\} \oplus d(\Lambda^{p-1}) \oplus d^*(\Lambda^{p+1})$

(b) $\dim \{\text{harmonic p-forms}\} < \infty$ and $\dim H^p(M;R) < \infty$

(c) Each cohomology class in the DeRham cohomology has a unique harmonic representative

(d) $\{\text{harmonic p-forms}\} \subseteq C^\infty(\Lambda^p)$.

Corollary: If d is exterior differentiation on the DeRham complex, then index $(d) = \chi(M)$, the Euler-Poincaré characteristic of M.

Proof:
$$\text{Index }(d) = \sum_{i=0}^{N} (-1)^i \dim N(D_i) \qquad \text{by definition}$$
$$= \sum_{i=0}^{N} (-1)^i \dim H^i(M,C)$$
$$= \chi(M).$$

Poincaré duality.

As a consequence of Hodge's theorem, we get Poincaré duality on oriented manifolds.

Let M be an orientable manifold with the orientation volume form dvol. Define the <u>Hodge star operator</u> $*:\Lambda^p T^*M \to \Lambda^{N-p} T^*M$ as the unique operator such that

$$(*\alpha,\beta)\text{dvol} = \alpha\wedge\beta \ .\text{dvol for } \alpha \in \Lambda^p \ , \ \beta \in \Lambda^{N-p} \ .$$

Alternately one can define $*$ by choosing a positively oriented orthonormal basis $\theta_1,\ldots,\theta_N$ for T^*M and setting

$$*\theta_{i_1}\wedge\cdots\wedge\theta_{i_p} = \varepsilon(i_1\ldots i_N)\theta_{i_{p+1}}\wedge\cdots\wedge\theta_{i_N}$$

where $\varepsilon(i_1\ldots i_N)$ is the sign of the permutation

$$\begin{pmatrix} 1 & 2 & \ldots & N \\ i_1 & i_2 & \ldots & i_N \end{pmatrix}$$

and extending $*$ by linearity to all of $\Lambda^p T^*M$.

<u>Proposition</u>.

(1) $** = (-1)^{p(N-p)}$

(2) $d^* = (-1)^{Np+N+1} * d *$ $\qquad d^*: \Lambda^p \to \Lambda^{p-1}$.

Proof: (1) It is sufficient to prove the equality on basis elements

$$**\theta_{i_1}\wedge\cdots\wedge\theta_{i_p} = *\varepsilon(i_1\ldots i_N)\theta_{i_{p+1}}\wedge\cdots\wedge\theta_{i_N}$$

$$= \varepsilon(i_1\ldots i_N)\varepsilon(i_{p+1},\ldots,i_N,i_1,\ldots i_p)\theta_{i_1}\wedge\cdots\wedge\theta_{i_p}$$

$$= (-1)^{p(N-p)}\theta_{i_1}\wedge\cdots\wedge\theta_{i_p}$$

(2) Let $\alpha \in \Lambda^{p-1}$ and $\beta \in \Lambda^{p}$. By Stokes theorem:

$$0 = \int_M d(*\beta\wedge\alpha) = \int_M d*\beta\wedge\alpha + (-1)^{N-p}*\beta\wedge d\alpha$$

$$= (*d*\beta,\alpha) + (-1)^{N-p}(**\beta,d\alpha)$$

$$= (*d*\beta,\alpha) + (-1)^{N-p+(N-p)p}(\beta,d\alpha)$$

$$= (*d*\beta,\alpha) + (-1)^{(N-p)(p+1)}(\beta,d\alpha).$$

Therefore:

$$(d\alpha,\beta) = (\alpha,*d*\beta)(-1)^{(N-p)(p+1)+1}$$

Since $(N-p)(p+1)+1 \equiv Np+N+1\ (2)$, this implies

$$d^* = (-1)^{Np+N+1}*d*.$$

Proposition: Let $D_p = d_{p-1}d^*_{p-1} + d^*_p d_p : \Lambda^p T^*M \to \Lambda^p T^*M$. Then $*D_p = D_p *$

Proof:

$$*D = *(dd^* + d^*d) = *(d*d*)(-1)^{N+Np+1} + (**d*d)(-1)^{N+1+N(p+1)}$$

$$D* = (dd^* + d^*d)* = (d*d**)(-1)^{N+N(N-p)+1} + (*d*d*)(-1)^{N+1+N(N-p+1)}.$$

But $N + Np + 1 \equiv N + 1 + N^2 - Np + N \pmod 2$

and

$$N + N^2 - Np + 1 + N(N - p) \equiv N + (+Np + N + p(N-p) \pmod 2 \Rightarrow *D = D*.$$

Corollary: If M is oriented, then $\dim H^p(M,\mathbb{C}) = \dim H^{N-p}(M,\mathbb{C})$.

Proof. $*D = D*$ and $** = (-1)^{p(N-p)}$ yield

$$*: N(D_p) \to N(D_{N-p}) \quad \text{1-1 and onto.}$$

but $H^p(M,C) \leftrightarrow N(D_p)$ (bijective correspondence)

and $H^{N-p}(M,C) \leftrightarrow N(D_{N-p})$.

Corollary: M oriented and odd dimensional $\Rightarrow \chi(M) = 0$.

Proof. Follows from the previous result.

Chapter III

Local Invariants Depending On The Eigenvalues of An Elliptic Operator

Let E be an Hermitian vector bundle over M. We showed in the last section that if $D:C^\infty(E) \to C^\infty(E)$ is an elliptic self-adjoint operator of positive order then

$$L^2(E) = N(D) \oplus R(D) \quad \text{and} \quad N(D) \subseteq C^\infty(E).$$

Now D is elliptic $\Rightarrow D - \lambda I$ is elliptic so that $(D-\lambda I)\phi = 0 \Rightarrow \phi \in C^\infty(E)$ i.e. the eigenfunctions of an elliptic operator are smooth. $D:L^2(E) \to L^2(E)$ is an unbounded operator with discrete spectrum $\{\lambda_n\}$ of eigenvalues with $0 \leqslant \lambda_n \to +\infty$. The associated eigenfunctions $\{\phi_n\}$, which can be assumed to be orthonormal with respect to the L^2 norm, form a basis for $L^2(E)$ and in fact $\{\phi_n\} \subseteq C^\infty(E)$. We say that $\{\lambda_n, \phi_n\}$ is a <u>spectral decomposition</u> of D.

These facts from the theory of unbounded operators will not be proved but we will look at some simple examples to make the above assertions plausible.

<u>Example:</u>

$$S^1 = \{z \in \mathbb{C}: |z| = 1\}. \quad \text{Let} \quad D = -\frac{d^2}{dx^2}, \quad D:L^2(S^1) \to L^2(S^1)$$

(x = coordinate function on S^1) .

D in this case has the spectral decomposition $\{n^2, e^{inx}\}_{n=-\infty}^{\infty}$. Observe that $\{n^2\}$ is a discrete subset of R^+, $\{n^2\} \to +\infty$. It is a well-known fact of classical Fourier Analysis that $\{e^{inx}\}$ form an orthonormal basis for $L^2(S^1)$. Clearly $\{e^{inx}\} \subseteq C^\infty(S^1)$.

$f \in L^2(S^1) \Rightarrow f(x) = \sum a_n e^{inx}$ where a_n are the Fourier coefficients of f.

The functional calculus for unbounded operators.

In the theory of complex variables one has Cauchy's integral formula

$$f(z) = \frac{1}{2\pi i} \int_C \frac{f(\xi)}{\xi - z} \, d\xi$$

where $f(z)$ is analytic in a neighborhood of the closed curve C and z is inside C.

For an unbounded operator $D: L^2(E) \to L^2(E)$ we define

$$f(D) = \frac{1}{2\pi i} \int_C \frac{f(\xi)}{(\xi I - D)} \, d\xi$$

where C is a curve about spec D and f is analytic and bounded in a neighborhood of spec D.

In all the cases we'll be interested in spec $D \subseteq R^+$, so that we can take C of the form

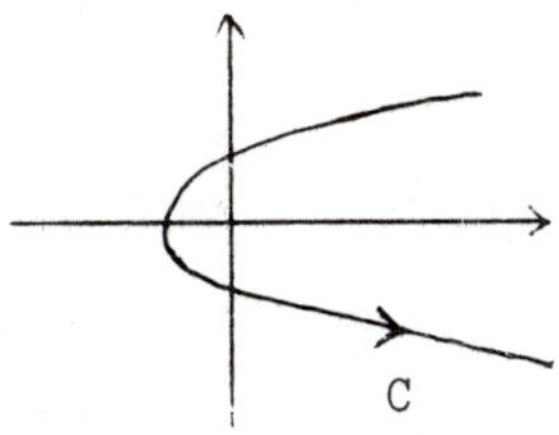

for suitable f (think of C as a curve on the Riemann sphere passing through the North Pole).

If $\{\lambda_n, \phi_n\}$ is a spectral decomposition of D we would expect that $f(D)\phi_n = f(\lambda_n)\phi_n$ and indeed

$$f(D)\phi_n = \frac{1}{2\pi i}\int_C \frac{f(\xi)}{\xi I - D} d\xi \quad \phi_n$$

$$= \frac{1}{2\pi i}\int_C f(\xi)(\xi I - D)^{-1}\phi_n d\xi .$$

But $(\xi I-D)\phi_n = \xi\phi_n - D\phi_n = \xi\phi_n - \lambda_n\phi_n = (\xi-\lambda_n)\phi_n \Rightarrow (\xi I-D)^{-1}\phi_n = (\xi-\lambda_n)^{-1}\phi_n$.

Therefore $f(D)\phi_n = \frac{1}{2\pi i}\int_C \left(f(\xi)(\xi-\lambda_n)^{-1}d\xi\right)\phi_n = f(\lambda_n)\phi_n$.

Example:

We return to our example of $-\frac{d^2}{dx^2} : L^2(S^1) \to L^2(S^1)$. Let $f_t(z) = e^{-tz}$ for $t \in \mathbb{C}$, $\operatorname{Re} t > 0$.

Then $e^{-tD} = \frac{1}{2\pi i}\int_C e^{-t\xi}(\xi I-D)^{-1}d\xi$ is well-defined for C of the form :

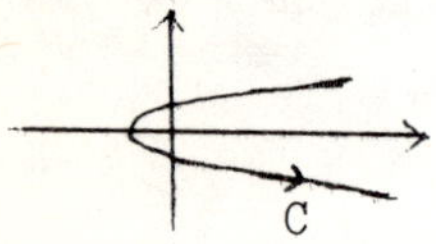

The integral converges absolutely because $e^{-t\xi}$ dies very quickly as $\xi \to \infty$ for $\operatorname{Re} t > 0$. We claim that e^{-tD} is an infinitely smoothing operator. D has spectral decomposition $\{n^2, e^{inX}\}_{n=-\infty}^{\infty}$
$(e^{-tD})e^{inx} = e^{-n^2t}\cdot e^{inx}$ so for $\phi \in L^2(S^1)$, $\phi(x) = \sum a_n e^{inx}$
$(e^{-tD}\phi)(x) = \sum a_n e^{-n^2t}e^{inx}$.

Since $\Sigma|n^r e^{-n^2 t} e^{inx}| < \infty$ for any r it follows that

$$e^{-tD}\phi \in H_s(S^1) \text{ for all } s, \text{ i.e. } e^{-tD}\phi \in C^\infty(S^1).$$

In particular $e^{-tD}: L^2(S^1) \to L^2(S^1)$ is bounded. It is a fact that infinitely smoothing operators have smooth kernel functions; we find one for e^{-tD}. We want to write $(e^{-tD}\phi)(x) = \int K(t,x,y)\phi(y)dy$ for some kernel $K(t,x,y)$.

Now

$$(e^{-tD}\phi)(x) = \sum a_n e^{-n^2 t} e^{inx}$$

$$= \sum\left(\int\phi(y)e^{-iny}dy\right)e^{-n^2 t}e^{inx}$$

$$= \int\left(\sum e^{-n^2 t}e^{in(x-y)}\right)\phi(y)dy.$$

Thus $K(t,x,y) = \sum e^{-n^2 t}e^{in(x-y)} \in C^\infty(S^1 \times S^1)$ for fixed t.

Observe that $\sum e^{-n^2 t}e^{in(x-y)}$ converges because of the factor $e^{-n^2 t}$.

It is well-known that $K(t,x,x) = \sum\limits_n e^{-n^2 t} \sim \frac{1}{\sqrt{4\pi t}}$ as $t \to 0^+$.

<u>Example</u>:

$$D = -\frac{d^2}{dx^2} : L^2(S^1) \to L^2(S^1).$$

This time take $f_s(z) = z^s$; $s \in \mathbb{C}$, $\operatorname{Re} s < -\frac{1}{2}$

$$f_s(D) = D^s = \frac{1}{2\pi i}\int_C \xi^s(\xi I - D)^{-1}d\xi$$

For $\phi \in L^2(S^1)$, $\phi(x) = \sum a_n e^{inx}$ we have

$$(D^s\phi)x = \sum a_n n^{2s} e^{inx}, \qquad \mathrm{Re}\ s < -\tfrac{1}{2}$$

$$= \sum \left(\int \phi(y) e^{-iny} dy \right) n^{2s} e^{inx}$$

$$= \int \left(\sum n^{2s} e^{in(x-y)} \right) \phi(y)\,dy$$

so that D^s has kernel function $K(s,x,y) = \sum_{n=0}^{\infty} n^{2s} e^{in(x-y)}$. However, in this case D^s is not an infinitely smoothing operator and $K(s,x,y) \in C^k(S^1 \times S^1)$ for $s < -\frac{1}{2} - k$, $k \in \mathbb{Z}^+$. Observe that $K(s,x,x) = \sum_{n=0}^{\infty} n^{2s}$ which is closely related to the Riemann Zeta Function. Clearly $K(s,x,x)$ blows up as $s \to -\frac{1}{2}$ and $K(s,x,x)$ has a meromorphic continuation to all of $\mathbb{C}$ with a simple pole at $s = -\frac{1}{2}$.

Kernel functions of elliptic operators on manifolds.

Let E be a Hermitian vector bundle over M and $D: L^2(E) \to L^2(E)$ a self-adjoint, positive elliptic operator. We can define

$$e^{-tD} = \frac{1}{2\pi i} \int_C e^{-t\xi}(\xi I - D)^{-1} d\xi \qquad \text{for} \quad \operatorname{Re} t > 0,$$

suitable C around spec D. Let $\{\lambda_n, \phi_n\}$ be a spectral decomposition for D. Then we have for $\phi \in L^2(E)$,

$$\phi = \sum_{n=0}^{\infty} a_n \phi_n$$

$$(e^{-tD}\phi)x = \sum_{n=0}^{\infty} a_n e^{-\lambda_n t} \phi_n(x) = \sum_{n=0}^{\infty} \int (\phi, \phi_n)_y dy \, e^{-\lambda_n t} \phi_n(x)$$

$$= \int \Big(\sum_{n=0}^{\infty} e^{-\lambda_n t} \phi_n(x) \otimes \overline{\phi_n(y)} \Big) \phi(y) dy$$

so that e^{-tD} has kernel function

$$K(t,x,y) = \sum_{n=0}^{\infty} e^{-\lambda_n t} \phi_n(x) \otimes \overline{\phi_n(y)}.$$

View $K(t,x,y)$ as an element of $\operatorname{Hom}(E_y, E_x)$. Then

$$\operatorname{Trace} K(t,x,x) = \sum_{n=0}^{\infty} e^{-\lambda_n t} (\phi_n, \phi_n)(x)$$

We shall show that e^{-tD} gives rise to local invariants depending only on Spec D by exhibiting an asymptotic expansion of Trace $K(t,x,x)$. Originally the problem of finding local invariants depending only on Spec D was tackled by R. Seeley who used the kernel $K(s,x,y)$ of the operator D^s. In this case.

$$\text{Trace } K(s,x,x) = \sum_{n=0}^{\infty} \lambda_n^s(\phi_n,\phi_n)_x \quad \text{for} \quad \operatorname{Re} s << 0$$

which has a meromorphic extension with isolated simple poles where $s = -\frac{\dim M}{2} + j$, $j = 0,1,2,\ldots,$ is a nonnegative integer. The values of Trace $K(s,x,x)$ at $s = -\frac{\dim M}{2} + j$, $j = 0,1,2,\ldots,$ can be computed using local formulae and are local invariants of D depending only on Spec D.

However, for technical reasons which will become clear later, we shall restrict our attention to the kernel $K(t,x,y)$ of the operator e^{-tD}, a procedure quite equivalent to using $K(s,x,y)$ since e^{-tD} and D^s are related by the

Mellin Transformation

$$\int_0^{\infty} e^{-tD} t^{s-1} dt = \Gamma(s)\, D^{-s} .$$

Proof: Let $f_s(x) = \int_0^\infty e^{-tx} t^{s-1} dt, \quad s << 0.$

Change coordinates by setting. $\tau = tx.$

Then $f_s(x) = \int_0^\infty e^{-\tau}\left(\frac{\tau}{x}\right)^{s-1} \frac{d\tau}{x} = \int_0^\infty e^{-\tau}\tau^{s-1}d\tau \cdot x^{-s} = \Gamma(s)\cdot x^{-s}$

Replacing x by D we get $\int_0^\infty e^{-tD} t^{s-1} dt = \Gamma(s)\cdot D^{-s}.$

What follows is the main theorem of this section.

Theorem: Let $D: L^2(E) \to L^2(E)$ be a positive self-adjoint elliptic operator of order 2. Then for $\operatorname{Re} t > 0$

(1) $e^{-tD}: L^2(E) \to L^2(E)$ is bounded

(2) $e^{-tD}: H_s(E) \to H_{s'}(E)$ is bounded for all s,s', hence is an infinitely smoothing operator

(3) e^{-tD} has a unique kernel function

$K(t,x,y) = \Sigma e^{-\lambda_n t} \phi_n(x) \otimes \bar{\phi}_n(y)$, analytic in t and C^∞ in x and y, such that

$$(e^{-tD}\phi)x = \int K(t,x,y)\phi(y)dy$$

(4) Trace $K(t,x,x) \sim \sum_{n=0}^{\infty} B_n(x,D)t^{(n-\dim M)/2}$ as $t \to 0^+$

(5) $B_n(x,D) \equiv 0$ for n odd and for n even can be computed invariantly in any local system as a C^∞ function of x, σD and the derivatives of σD.

For the purposes of the proof we change the definition of e^{-tD} slightly. Let $e^{-tD} = \frac{1}{2\pi i}\int_C e^{-t\lambda}(D-\lambda I)^{-1}d\lambda$ where C is oriented <u>clock-wise</u>.

<u>Definition</u>. For P a ψDO on M define $\|P\|_{s,s'} = \sup_{u\in H_{s'}} \frac{\|Pu\|_s}{\|u\|_{s'}}$

Our method of attack is to get a "good" approximation D_λ to $(D-\lambda I)^{-1}$ and set $E(t) = \frac{1}{2\pi i}\int_C e^{-t\lambda}D_\lambda d\lambda$. D_λ will be chosen so that $\|E(t) - e^{-tD}\|_{s,s'} \leq \varepsilon\cdot|t|^n$. We then show that $E(t)$ is infinitely smoothing; it must have a smooth kernel $L(t,x,y)$ which will satisfy $|L(t,x,y) - K(t,x,y)| \leq \varepsilon|t|^n$. It will be sufficient to prove everything for the approximations $E(t)$ and $L(t,x,y)$. To handle D_λ we need to define the <u>algebra of ψDO's depending on a complex paramater</u>.

Let $S^m_\lambda = \{p(x,\xi,\lambda)$: p is C^∞ in x and ξ, has compact x support, analytic in λ for $\lambda \in \mathbb{C} - \mathbb{R}^+$ and $|\partial^\alpha_\xi D^\beta_x p| \leq C_{\alpha\beta}(1+|\xi|+|\lambda|^{.5})^{m-|\alpha|}\}$. P_λ is said to be a <u>ψDO of order m depending on λ</u> if $\sigma P_\lambda = p \in S^m_\lambda$.

Also extend the definition of $\sim$ by setting $P_\lambda \sim 0$ if $\sigma P_\lambda \in S^m_\lambda$ for all m. The results we had for ordinary ψDO's extend to this new class of ψDO's.

<u>Proposition</u>:

(1) $\sigma P \in S^m_\lambda$ and $\sigma Q \in S^{m'}_\lambda \Rightarrow \sigma(PQ) \in S^{m+m'}_\lambda$ and $\sigma(PQ) \sim \sum_\alpha \partial^\alpha_\xi p D^\alpha_x q/\alpha!$

(2) $\sigma P \in S^m_\lambda \Rightarrow \sigma P^* \in S^m_\lambda$ and $\sigma P^* \sim \sum_\alpha \partial^\alpha_\xi D^\alpha_x p^*/\alpha!$

(3) The leading order symbols of elements in S^m_λ are invariantly defined on the cotangent bundle.

We shall omit the proof of this proposition since it is essentially the same as in the ordinary case. As before, everything can be extended to manifolds by partitions of unity.

We also have the following estimate which will be useful in the proof of the main theorem.

Lemma: Given $k \in \mathbf{Z}^+$ and $s,s' \in \mathbb{R}$. Then $\exists m << 0$ such that $p \in S^m_\lambda \Rightarrow \|P_\lambda\|_{s,s'} \leqslant C(1+|\lambda|)^{-k}$ where $m = m(s,s',k)$.

Proof: essentially the same as in the ordinary case.

Proof of main theorem: Since D is 2nd order, $\sigma D = a_2(x,\xi) + a_1(x,\xi) + a_0(x,\xi)$ where the a_i are matrices of polynomials of order i.

D self-adjoint & positive $\Rightarrow A_2$ self-adjoint & positive where $\sigma A_2 = a_2$

$$\Rightarrow \text{Spec } a_2 \subseteq \mathbb{R}^+$$

$$\Rightarrow (a_2(x,\xi)-\lambda)^{-1} \text{ exists provided } \lambda \notin \mathbb{R}^+.$$

Our first objective is to invert the elliptic operator $(D-\lambda I)$ which is a ψDO depending on a complex parameter. We proceed analogously to the way in which an ordinary elliptic operator is inverted.

Suppose B_λ is an inverse to $D-\lambda I$ and $\sigma B_\lambda = b_0 + b_1 + \ldots$ with $b_k \in S^{-2-k}_\lambda$.

Define $b_0 = (a_2(x,\xi)-\lambda)^{-1} \in S^{-2}_\lambda$ for $\lambda \notin \mathbb{R}^+$ since A_2 is a differential operator

and
$$b_k = -b_0 \sum_{\substack{2+|\alpha|+j-i=k \\ j<k}} \partial^\alpha_\xi b_j D^\alpha_x a_i/\alpha! \in S^{-2-k}_\lambda .$$

(We want $b_k \in S_\lambda^{-2-k}$ so we must have $-k = -2 - j - |\alpha| + i$ which explains why the summation is over $2 + |\alpha| + j - i = k$). Now let s, s' and n be given. Choose N_o so $\sigma(D_\lambda) = \sum_{k=0}^{N_o}$ satisfies

$$\sigma((D_\lambda(D-\lambda I))-I) \in S_\lambda^{-m} \quad \text{with } m \text{ chosen so that}$$

$$\|D_\lambda(D-\lambda I)-I\|_{s,s'} \leq C(1+|\lambda|)^{-n+1}$$

$$\Rightarrow \|D_\lambda-(D-\lambda I)^{-1}\|_{s,s'} \leq C(1+|\lambda|)^{-n+1} .$$

Now <u>define</u> $E(t) = \frac{1}{2\pi i}\int_C e^{-t\lambda}D_\lambda d\lambda$, $\mathrm{Re}(t) > 0$.

Then $\|E(t) - e^{-tD}\|_{s,s'} \leq \frac{1}{2\pi}\int_C e^{-t\lambda}\|D_\lambda-(D-\lambda I)^{-1}\|_{s,s'}d\lambda$. Let $\lambda' = t\lambda$,

$$\leq \frac{1}{2\pi}\int_C e^{-\lambda'}\|D_{\frac{\lambda'}{t}} - (D-\frac{\lambda'}{t}I)^{-1}\|_{s,s'}\frac{d\lambda'}{t} \leq \frac{1}{2\pi}\int_C e^{-\lambda'}|\frac{\lambda'}{t}|^{-n+1}\frac{d\lambda'}{t}$$

$\leq \varepsilon|t|^{-n}$ for some constant ε.

For the remainder of the proof we make the assumption that $a_2(x,\xi) = a_2(x,\xi)\, I.$, i.e. a_2 is a scalar times the identity matrix. All operators of interest to us satisfy this condition e.g. if D is the Laplacian, $\sigma_L D = |\xi|^2 I$. Under this assumption, $b_k = \sum_j b_{k,j}(x,\xi)(a_2-\lambda)^{-j}$ since everything commutes.

Let $$e_k(x,\xi) = \frac{1}{2\pi i}\int_C b_k(x,\xi,\lambda)e^{-t\lambda}d\lambda$$

$$= \sum_j b_{k,j}(x,\xi)\,\frac{1}{2\pi i}\int_C e^{-t\lambda}(a_2-\lambda)^{-j}d\lambda$$

$$= \sum_j b_{k,j}(x,\xi)\big/(j-1)!\,t^{j-1}e^{-ta_2(x,\xi)} \quad \text{by the Residue Theorem.}$$

Now $e_k(x,\xi) \in S^{-m}$ for all m since $e^{-ta_2(x,\xi)}$ dies very rapidly near infinity for $\operatorname{Re} t > 0$. It is clear that

$$\sigma(E(t)) = \sum_{k=0}^{N_0} e_k(x,\xi)$$

$$= \sum_{k=0}^{N_0} \sum_j b_{k,j}(x,\xi)\big/(j-1)!\,t^{j-1}e^{-ta_2(x,\xi)}$$

and $\sigma(E(t)) \in S^{-m}$ for all m. Thus $E(t)$ is infinitely smoothing, therefore $E(t)$ is a ψDO $\Rightarrow e^{-tD}$ is infinitely smoothing

$\Rightarrow e^{-tD}: L^2(E) \to L^2(E)$ is bounded.

Now $(E(t)\phi)x = \int e^{i(x-y)\cdot\xi}\sigma(E(t))dyd\xi.$

Let $L(t,x,y) = \int e^{i(x-y)\cdot\xi}\sigma(E(t))d\xi$, which is well-defined since $\sigma(E(t)) \in S^{-m}$ for $m > \dim M/2$.

Since $\sigma(E(t)) \in S^{-m-k}$ for $m > \dim M/2$ and all k we can differentiate $L(t,x,y)$ as often as we want so that $L(t,x,y) \in C^\infty$. Thus $E(t)$ has the smooth kernel function $L(t,x,y)$ which implies that $K(t,x,y)$ is a smooth kernel function for e^{-tD}.

Now $L(t,x,x) = \int (E(t))d\xi$

$$= \sum_{k=0}^{n_0} \sum_j \int b_{k,j}(x,\xi)/(j-1)!t^{j-1}e^{-ta_2(x,\xi)} d\xi .$$

Make the change of variables $\eta = \xi\sqrt{t}$.

Then $d\xi = d(\eta/\sqrt{t}) = t^{-\dim M/2} d\eta$.

$$a_2(x,\xi) = a_2(x,\eta/\sqrt{t}) = t^{-1} a_2(x,\eta)$$

and

$$b_{k,j}(x,\xi) = b_{k,j}(x,\eta/\sqrt{t}) = t^{(2+k-2j)/2} b_{k,j}(x,\eta).$$

Therefore

$$L(t,x,x) = \sum_{k=0}^{N_0} \sum_j \int b_{k,j}(x,\eta)/(j-1)!e^{-a_2(x,\eta)} t^{(2+k-2j)/2} t^{j-1-\dim M/2} d\eta$$

$$= \sum_{k,j} \int b_{k,j}(x,\eta)e^{-a_2(x,\eta)} d\eta \cdot t^{(k-\dim M)/2}/(j-1)!$$

Let $B_k(x,D) = \text{Trace} \sum_j \int b_{k,j}(x,\eta)e^{-a_2(x,\eta)} d\eta/(j-1)!$

Then $\text{Trace } L(t,x,x) = \sum_{k=0}^{N_0} B_k(x,D)t^{(k-\dim M)/2}$

which proves that $\text{Trace } K(t,x,x) \sim \sum_{n=0}^{\infty} B_n(x,D)t^{(n-\dim M)/2}$ and $B_n(x,D)$ are computable in terms of $\sigma(D)$ and its derivatives.

Lastly, observe that since order $b_{k,j} = -2 - k + 2j$ we have that order $b_{k,j} \equiv k \pmod 2$. Thus k odd $\Rightarrow b_{k,j}(x,\eta)$ is an odd function of η $\Rightarrow \int b_{k,j}(x,\eta)e^{-a_2(x,\eta)} d\eta = 0$

so that $B_n(x,D) = 0$ when n is odd.

Let $D\colon C^\infty(M) \to C^\infty(M)$ be the Laplacian on functions, $(x_1,\ldots,x_N)$ local coordinates and G the Riemannian metric on M.

$$G\left(\frac{\partial}{\partial x_i}, \frac{\partial}{\partial x_j}\right)(x) = g_{ij}(x)$$

$$g = \det(g_{ij}) \quad , \qquad (g^{ij}) = (g_{ij})^{-1} .$$

Then $D = -\sum_{i,j} \frac{1}{\sqrt{g}} \frac{\partial}{\partial x_i}\left(g^{ij}\sqrt{g}\,\frac{\partial}{\partial x_j}\right)$.

It is easy to see that

$$\sigma D = a_2(x,\xi) + a_1(x,\xi) + a_0(x,\xi) \quad \text{with}$$

$$a_2(x,\xi) = \sum_{i,j} g^{ij}\xi_i\xi_j$$

$$a_1(x,\xi) = \sum_{i,j} (\text{expression involving } 1^{st} \text{ order derivatives of } (g_{ij}))\cdot\xi_i ,$$

$$a_0(x,\xi) = \sum_{i,j} (\text{expression involving } 2^{nd} \text{ order derivatives of } (g_{ij})).$$

Thus in terms of derivatives of the metric $a_i(x,\xi)$ has "degree" 2-i. Recall that $b_k = -b_0 \sum_{\substack{i,j,\alpha \\ 2-i+|\alpha|+j=k}} \partial_\xi^\alpha b_j D_x^\alpha a_i/\alpha!$.

Now b_0 has degree 0 in terms of derivatives of the metric. Assume inductively that b_j has degree j in terms of derivatives of the metric, for $j < k$. Then each term in b_k has degree $j + 2 - i + |\alpha| = k$

in terms of derivatives of the metric $\Rightarrow B_n(x,D)$ is a homogeneous polynomial of degree n in the derivatives of the metric, since B_n is obtained from b_n by integrating the latter in the ξ-direction. Though we shall not prove it, this result is true for general D. Thus to the main theorem we have the

<u>Corollary</u>: $B_n(x,D)$ is a homogeneous polynomial of degree n in the derivatives of the symbol of D.

An Index Theorem for Elliptic Complexes.

We showed that for any second-order, positive, self-adjoint elliptic operator $D:C^\infty(E) \to C^\infty(E)$

$$f(t,x,D) = \text{Trace } K(t,x,x)$$

$$= \sum_{n=0}^{\infty} e^{-\lambda_n t}(\phi_n,\phi_n)x \sim \sum_{n=0}^{\infty} B_n(x,D)t^{(n-m)/2} \quad \text{as} \quad t \to 0^+$$

where m = dimension M.

Since $\{\phi_n\}$ are chosen to be an orthonormal basis, this implies

$$\int_M f(t,x,D)\,dvol = \sum_{n=0}^{\infty} e^{-\lambda_n t} \sim \sum_{n=0}^{\infty} \left(\int_M B_n(x,D)\,dvol\right)t^{(n-m)/2} \quad .$$

Now let $d_i: C^\infty(E_i) \to C^\infty(E_{i+1})$ be an elliptic complex over M, dimension $M = m$. Construct the Laplacian

$$D_i = d_{i-1}d_{i-1}^* + d_i^*d_i : C^\infty(E_i) \to C^\infty(E_i) \qquad \text{and define}$$

$$B_n(x,d) = \sum_{i=0}^{n} (-1)^i\, B_n(x,D_i).$$

By our previous results $B_n(x,d)$ is computable by means of local formulae. The central role played by this invariant is indicated in the next theorem.

<u>Theorem</u>: $\int_M B_n(x,d)\,dvol = \begin{cases} 0, \text{ when } n \neq m \\ \text{index }(d), \text{ when } n = m . \end{cases}$

<u>Lemma</u>: Let E_i^λ be the eigenspace corresponding to λ of the operator D_i. Then for $\lambda \neq 0$, the sequence

$$0 \to E_0^\lambda \xrightarrow{d_0} E_1^\lambda \xrightarrow{d_1} \cdots \xrightarrow{d_{n-1}} E_n^\lambda \to 0$$

is exact.

<u>Proof</u>: Firstly, observe that the sequence is defined: for if $\phi \in E_i^\lambda$, then $D_{i+1}(d_i\phi) = d_iD_i\phi = d_i\lambda\phi = \lambda(d_i\phi)$. To show the sequence is exact, suppose $d_i\phi = 0$ for $\phi \in E_i^\lambda$, $i = 1,\ldots,n-1$. Then

$$\lambda\phi = D_i\phi = (d_{i-1}d_{i-1}^* + d_i^*d_i)\phi = d_{i-1}d_{i-1}^*\phi$$

so that $\phi = d_{i-1}\left(\frac{d_{i-1}^*\phi}{\lambda}\right)$. But $\frac{d_{i-1}^*\phi}{\lambda} \in E_{i-1}^\lambda$

since $D_{i-1}\left(\frac{d_{i-1}^*\phi}{\lambda}\right) = (d_{i-2}d_{i-2}^* + d_{i-1}^*d_{i-1})\left(\frac{d_{i-1}^*\phi}{\lambda}\right)$

$= \frac{1}{\lambda}(d_{i-1}^*d_{i-1}d_{i-1}^*\phi) = \frac{1}{\lambda}(d_{i-1}^*(d_{i-1}d_{i-1}^*+d_i^*d_i)\phi) = \frac{1}{\lambda}(d_{i-1}^*\lambda\phi) = d_{i-1}^*\phi.$

Similarly, $d_0\colon E_0^\lambda \to E_1^\lambda$ is injective and

$d_{n-1}\colon E_{n-1}^\lambda \to E_n^\lambda$ is surjective.

Proof of Theorem: Define $f(t,x,d) = \sum_{i=0}^{n} (-1)^i f(t,x,D_i)$.

Then $\int_M f(t,x,d)dvol = \sum_{i=0}^{n} (-1)^i \int_M f(t,x,D_i)dvol$

$= \sum_{i=0}^{n} (-1)^i \sum_{\lambda} e^{-\lambda t} \dim E_i^{\lambda}$ (the term $\dim E_i^{\lambda}$ takes into account the multiplicity of the eigenvalue λ)

$= \sum_{\lambda} e^{-\lambda t} \sum_{i=0}^{n} (-1)^i \dim E_i^{\lambda}$

Now it follows from the lemma that for $\lambda \neq 0$ $\sum_{i=0}^{n} (-1)^i \dim E_i^{\lambda} = 0$.

Therefore

$$\int_M f(t,x,d)dvol = \sum_{i=0}^{n} (-1)^i \dim E_i^0 = \sum_{i=0}^{n} (-1)^i \ker D_i = \text{index } (d) .$$

But we also have

$$\int_M f(t,x,d)dvol = \sum_{i=0}^{n} (-1)^i \int_M f(t,x,D_i)dvol$$

$$\sim \sum_{i=0}^{n} (-1)^i \sum_{n=0}^{\infty} (\int_M B_n(x,D_i)dvol)t^{(n-m)/2} \quad \text{as} \quad t \to 0^+$$

$$= \sum_{n=0}^{\infty} (\int_M B_n(x,d)dvol)t^{(n-m)/2}$$

Thus $\text{index}(d) \sim \sum_{n=0}^{\infty} (\int_M B_n(x,d)dvol)t^{(n-m)/2}$.

But the left-hand side is completely independent of t. The only possibility, therefore, is

$$\int_M B_n(x,d)\,dvol = \begin{cases} 0 & \text{when } n \neq m \\ \text{index } (d) & \text{when } n = m \end{cases}$$

Corollary. Dimension M odd $\Rightarrow$ index $(d) = 0$.

The index theorem just proved is the starting point for proving the index theorems of the classical elliptic complexes, which we state now (definitions and proofs later).

(1) For the DeRham complex

index $(d) = \chi(M)$ = Euler characteristic of M

$B_m(x,d)$ = Euler class.

(2) For the Dolbeault complex of a Kahler manifold

index $(\bar{\partial})$ = arithmetic genus of M

$B_m(x,\bar{\partial})$ = Riemann-Roch-Hirzebruch polynomial in the Chern classes.

(3) For the signature complex of a 4k-dimensional orientable manifold

index $(d+d^*)$ = sign (M)

$B_m(x,d+d^*)$ = Hirzebruch's L_k polynomial.

(4) For the spin complex of a spin manifold

index $(d_s) = \hat{A}(M)$ = the spinor index of M

$B_m(x,d_s)$ = the $\hat{A}$- polynomial.

The major task in proving these results is to identity B_m with a suitable known invariant in each case.

Some easy consequences are:

Corollary: Let $\tilde{M}$ be an n-fold covering of M. The following hold when they make sense.

(1) $\chi(\tilde{M}) = n \cdot \chi(M)$

(2) The arithmetic genus of $\tilde{M}$ $= n\cdot$ the arithmetic genus of M.

(3) sign $(\tilde{M}) = n$ sign (M)

(4) $\hat{A}(\tilde{M}) = n \cdot \hat{A}(M)$.

Proof: The Riemannian metric on M pulls back under the projection $\pi: \tilde{M} \to M$ to a Riemannian metric on $\tilde{M}$, making π a local isometry. Thus e.g.

$$\chi(\tilde{M}) = \int_{\tilde{M}} B_m = n \cdot \int_M B_m = n\, \chi(M).$$

Corollary: Let M_1 and M_2 be manifolds of the same dimension and $M_1 \# M_2$ the connected sum of M_1 and M_2. Then

(1) $\chi(M_1 \# M_2) = \chi(M_1) + \chi(M_2) - 2$

(2) If $\dim M_1 = \dim M_2 = 4k$ for some k and both M_1 and M_2 are orientable, then

sign $(M_1 \# M_2) =$ sign $(M_1) +$ sign(M_2).

Proof: $$\int_{M_1 \# M_2} B_m + \int_{S^m} B_m = \int_{M_1} B_m + \int_{M_2} B_m.$$

Therefore $\chi(M_1 \# M_2) + \chi(S^m) = \chi(M_1) + \chi(M_2)$

$$\Rightarrow \chi(M_1 \# M_2) = \chi(M_1) + \chi(M_2) - 2$$

and sign $(M_1 \# M_2)$ + sign (S^m) = sign (M_1) + sign (M_2).

But sign $(S^m) = 0$

$\Rightarrow$ sign $(M_1 \# M_2)$ = sign (M_1) + sign (M_2).

Chapter IV

The Chern-Gauss-Bonnet Formula for $\chi(M)$

We first prove this theorem in dimension 2 to get the classical Gauss-Bonnet theorem $\chi(M^2) = \frac{1}{2\pi}\int_{M^2} K\, d\mathbf{vol}$; K is the scalar curvature of M^2. We have index $(d) = \chi(M^2) = \int_{M^2} B_2(x,d)dvol$.

It is sufficient to show that $B_2(x,d) = \lambda K$, λ some constant. Since the local formula for $B_2(x,d)$ works for <u>all</u> 2-manifolds

$$\int_{S^2}\lambda K = \chi(S^2) = 2 \Rightarrow \lambda = \frac{1}{2\pi} .$$

Now the metric tensor on M^2 is of the form

$$ds^2 = g_{11}dx_1^2 + 2g_{12}dx_1dx_2 + g_{22}dx_2^2 \quad \text{with} \quad g_{ij} = G\left(\frac{\partial}{\partial x_i}, \frac{\partial}{\partial x_j}\right),$$

G being a Riemannian metric on M^2.

We'll adopt the notation $g_{ij/\omega} = d_x^\omega g_{ij}$; ω a multi-index. The scalar curvature K has the following expression in terms of the metric and its derivatives

$$K = \frac{1}{\sqrt{g}}\left\{\frac{\partial}{\partial x_1}\left(\frac{g_{12}g_{11/2}-g_{11}g_{22/1}}{2g_{11}\sqrt{g}}\right) + \frac{\partial}{\partial x_2}\left(\frac{2g_{11}g_{12/1}-g_{12}g_{11/1}-g_{11}g_{11/2}}{2g_{11}\sqrt{g}}\right)\right\}$$

where $g = \det(g_{ij}) > 0$.

We normalize our coordinates $X=(x_1,x_2)$ at x_0 by requiring that $g_{ij}(x_0) = \delta_{ij}$ and $g_{ij/k}(x_0) = 0$. Geodesic polar coordinates are normalized, but there are many other such systems. For such normal coordinates K has the much simpler form $K = -\frac{1}{2}\{g_{11/22} + g_{22/11} - 2g_{12/12}\}$. Let $\mathcal{P}$ denote the ring of polynomials by the indeterminates $g_{ij/\omega}$. For coordinate chart X, metric G, define $g_{ij/\omega}(X,G)(x_0) = \partial_x^{\omega}(g_{ij})(x_0)$. For $P \in \mathcal{P}$, define $P(X,G)(x_0)$ analogously. $P \in \mathcal{P}$ is said to be invariant if $P(X,G)(x_0) = P(Y,G)(x_0)$ for all coordinate charts X and Y, i.e. the value of P at x_0 depends only on the metric G.

We shall consider only normal coordinate systems centered at x_0, so we can view $\mathcal{P}$ as a ring in the $g_{ij/\omega}$ variables with $|\omega| \geq 2$. If $A \in \mathcal{P}$ is a monomial, define order (A) to be the total number of derivatives of g_{ij} appearing in A.

For $P \in \mathcal{P}$, define order(P) = max {order (A): A a monomial of P}
$P \in \mathcal{P}$ is homogeneous if order (A) = order (P) for all monomials A of P.

Lemma: Let $P \in \mathcal{P}$ be such that $P(X,G)(x_0) = 0$ for all (X,G). Then $P \equiv 0$.

Proof: For any set of numbers $c_{ij,\omega}$ using Taylor's series we can construct a metric G with $g_{ij/\omega}(x_0) = c_{ij,\omega}$ Since we can do this for as many different sets $c_{ij,\omega}$ as we please, this implies $P \equiv 0$.

1st Proof of Gauss-Bonnet (Geometric):

We'll prove that if $P \in \mathcal{P}$ is homogeneous and invariant of order 2 then $P = \lambda K$ for some λ.

Choose λ so that $P-\lambda K$ does not contain the monomial $g_{22/11}$. It is a well-known fact of surface theory that given G, one can choose a coordinate system $X = (x_1, x_2)$ such that $ds^2 = dx_1^2 + g_{22}dx_2^2$

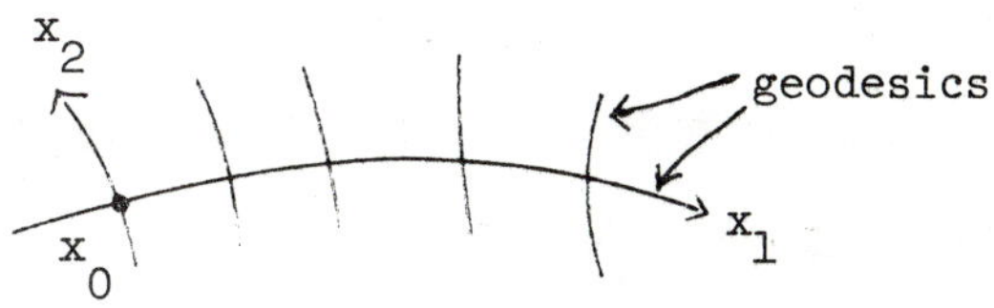

with $g_{22}(x_0) = 1$, $\quad g_{22/1}(x_0) = g_{22/2}(x_0) = 0$

$$g_{22/12}(x_0) = g_{22/22}(x_0) = 0 \quad .$$

Thus $(P-\lambda K)(X,G)(x_0) = 0$.

Since G was arbitrary and $P-\lambda K$ is invariant we must have

$$(P-\lambda K)(Y,G)(x_0) = 0 \quad \text{for } \underline{\text{all}} \ Y,G.$$

Hence by the lemma, $P-\lambda K \equiv 0$. So let $P = B_2$. Actually we proved more than Gauss-Bonnet, viz: if $L(P)$ denotes the coefficient of $g_{22/11}$ in an invariant homogeneous $P \in \mathcal{P}$ of order 2 and $L(P) = 0$, then $P \equiv 0$

$\Rightarrow$ dim {homogeneous invariants of order 2 on M^2} ≤ 1.

The existence of K

$\Rightarrow$ dim {homogeneous invariants of order 2 on M^2} = 1

$\Rightarrow$ Gauss-Bonnet.

The proof just given relies heavily on being able to find a very special coordinate system which cannot be found on higher dimensional manifolds. Therefore we give another proof which is combinatorial in nature and can be generalized to higher dimensions.

2nd Proof of Gauss-Bonnet (combinatorial):

Notation: If A is any monomial in the $g_{ij/\omega}$, $\deg_k A$ means the number of times the index k appears in A.

Let $X = (x_1, x_2)$ be any normal coordinate system. Choose the new coordinate system Y with $y_1 = -x_1$ $y_2 = x_2$. Then

$$g^Y_{ij/\omega} = (-1)^{\left(\deg_1 g^X_{ij/\omega}\right)} g^X_{ij/\omega} \qquad \text{e.g.} \quad g^Y_{12} = G\left(\frac{\partial}{\partial y_1}, \frac{\partial}{\partial y_2}\right) = G\left(-\frac{\partial}{\partial x_1}, \frac{\partial}{\partial x_2}\right)$$

$$= -g^X_{12}\,.$$

Now let $P \in \mathcal{P}$ be a 2nd order homogeneous invariant polynomial, $P = \Sigma c_P(A)A$ where $c_P(A)$ = the coefficient of the monomial A in P.

$$P(Y,G) = P(X,G) \Rightarrow \Sigma c_P(A)A = \Sigma c_P(A)(-1)^{\deg_1 A} A$$

$$\Rightarrow c_P(A) = (-1)^{\deg_1 A} c_P(A)$$

$$\Rightarrow c_P(A) = 0 \text{ if } \deg_1 A \text{ is odd.}$$

One can interchange the indices 1 and 2 in the above argument. Consequently P consists only of terms of even degree in both indices.

Now consider the new coordinates Y such that

$$y_1 = x_2, \quad y_2 = x_1$$

If $P = a_1 g^X_{11/11} + a_2 g^X_{22/22} + a_3 g^X_{11/22} + a_4 g^X_{22/11} + a_5 g^X_{12/12} + a_6 g^X_{21/21}$

then

$$P = a_1 g^Y_{22/22} + a_2 g^Y_{11/11} + a_3 g^Y_{22/11} + a_4 g^Y_{11/22} + a_5 g^Y_{21/21} + a_6 g^Y_{12/12}$$

$$\Rightarrow \quad a_1 = a_2, \quad a_3 = a_4, \quad a_5 = a_6 \, .$$

Thus we can restrict ourselves to polynomials of the type

$$P = \alpha(g_{11/11} + g_{22/22}) + \beta(g_{11/22} + g_{22/11}) + \gamma g_{12/12} \quad .$$

Next we show that $\alpha = 0$ by choosing a nonlinear change of coordinates Y such that

$$x_1 = y_1 + \varepsilon y_1^3 \, , \qquad x_2 = y_2 \quad \text{in which case}$$

$$\frac{\partial}{\partial y_1} = (1 + 3\varepsilon y_1^2)\frac{\partial}{\partial x_1} \quad \text{and} \quad \frac{\partial}{\partial y_2} = \frac{\partial}{\partial x_2} \quad .$$

We have $g^X_{ij} = \delta_{ij} + O(x)$. One easily computes that

$$g^Y_{11} = g^X_{11}(1 + 6\varepsilon y_1^2) + O(x^3)$$

$$g^Y_{12} = g^X_{12} + O(x^3)$$

$$g^Y_{22} = g^X_{22}$$

from which it follows that $g^Y_{11/11} = g^X_{11/11} + 12\varepsilon + O(x)$ while $g^Y_{ij/\omega} = g^X_{ij/\omega}$ for all other $g_{ij/\omega}$ in P.
Therefore

$$P(Y,G)(x_0) = [\alpha(g^Y_{11/11} + g^Y_{22/22}) + \beta(g^Y_{11/22} + g^Y_{22/11}) + \gamma g^Y_{12/12}]\,(x_0)$$

$$= 12\alpha\varepsilon + P(X,G)(x_0)$$

$$\Rightarrow \alpha = 0$$

$$\Rightarrow P = \beta(g_{11/22} + g_{22/11}) + \gamma g_{12/12}.$$

Finally, look at coordinate changes $Y = (y_1, y_2)$ such that

$$\frac{\partial}{\partial y_1} = a\frac{\partial}{\partial x_o} + b\frac{\partial}{\partial x_2} \qquad \text{with } a^2 + b^2 = 1$$

$$\frac{\partial}{\partial y_2} = -b\frac{\partial}{\partial x_1} + a\frac{\partial}{\partial x_2} \qquad \text{i.e. } \begin{pmatrix} a & b \\ -b & a \end{pmatrix} \in SO(2) \quad .$$

In computing $g^Y_{ij/\omega}$, formally $1 \mapsto a1 + b2$, $2 \mapsto -b1 + a2$.
Thus

$$\begin{cases} g^Y_{11} = a^2 g^X_{11} + 2ab g^X_{12} + b^2 g^X_{22} \\ g^Y_{22} = b^2 g^X_{11} - 2ab g^X_{12} + a^2 g^X_{22} \\ g^Y_{12} = -ab g^X_{11} + (a^2 - b^2) g^X_{12} + ab g^X_{22} \end{cases}$$

from which we get

$$g^Y_{11/22} = a^2b^2g^X_{11/11} + 2a^3bg^X_{11/12} + a^4g^X_{11/22} + \dots$$

$$g^Y_{22/11} = a^2b^2g^X_{11/11} + \dots$$

$$g^Y_{12/12} = a^2b^2g^X_{11/11} + \dots$$

Therefore $P(Y,G) = \beta(g^Y_{11/22} + g^Y_{22/11}) + \gamma g^Y_{12/12}$

$$= (2\beta + \gamma)a^2b^2\ (g^X_{11/11} + \dots)$$

$$= P(X,G)$$

$$= \beta(g^X_{11/22} + g^X_{22/11}) + \gamma g^X_{12/12}$$

$\Rightarrow (2\beta + \gamma)a^2b^2 = 0$

$\Rightarrow 2\beta + \gamma = 0$

$\Rightarrow P = \beta(g_{11/22} + g_{22/11}) - 2\beta g_{12/12}$

$$= \beta(g_{11/22} + g_{22/11} - 2g_{12/12})$$

$\Rightarrow$ dim {homogeneous invariant polynomials of order 2} $\leqslant 1$

$\Rightarrow$ $B_2 = \lambda K$ for some λ.

We will need the following functorial properties of the invariants $B_n(x,D)$.

Proposition:

(1) $B_n(x,aD) = a^{(n-m)/2} B_n(x,D)$, $a > 0$.

(2) $B_n(x,D+a) = \sum_{k=0}^{n/2} \frac{(-a)^k t^k}{k!} B_{n-2k}(x,D)$, $a > 0$.

(3) Let E and E' be Hermitian vector bundles over the same manifold M, and $D: C^\infty(E) \to C^\infty(E)$

$D': C^\infty(E') \to C^\infty(E')$

positive self-adjoint elliptic operators. Then

$B_n(x,D+D') = B_n(x,D) + B_n(x,D')$ where

$D + D': C^\infty(E) \oplus C^\infty(E') \to C^\infty(E) \oplus C^\infty(E')$.

Proof: (1) If $\{\lambda_n, \phi_n\}$ is a spectral decomposition of D, then $\{a\lambda_n, \phi_n\}$ is a spectral decomposition of aD. Therefore

$$\sum_{n=0}^{\infty} B_n(x,aD) t^{(n-m)/2} \sim f(t,x,aD) = \sum_{n=0}^{\infty} e^{-a\lambda_n t} (\phi_n, \phi_n)x$$

$$\sim \sum_{n=0}^{\infty} B_n(x,D)(at)^{(n-m)/2} = \sum_{n=0}^{\infty} a^{(n-m)/2} B_n(x,D) t^{(n-m)/2}$$

$\Rightarrow B_n(x,aD) = a^{(n-m)/2} B_n(x,D)$.

(2) If $\{\lambda_n, \phi_n\}$ is a spectral resolution of D, then $\{\lambda_n + a, \phi_n\}$ is a spectral resolution of $D + a$. Therefore

$$\sum_{n=0}^{\infty} B_n(x,D+a)t^{(n-m)2} \sim f(t,x,D+a) = \sum_{n=0}^{\infty} e^{-at}e^{-\lambda_n t}(\phi_n,\phi_n)x$$

$$= e^{-at}f(t,x,D) \sim \sum_{k=0}^{\infty} \frac{(-a)^k t^k}{k!} \qquad \sum_{n=0}^{\infty} B_n(x,D)t^{(n-m)/2}$$

$$\Rightarrow B_n(x,D+a) = \sum_{k=0}^{n/2} \frac{(-a)^k}{k!} B_{n-2k}(x,D).$$

(3) If $\{\lambda_n, \phi_n\}$ and $\{\lambda'_n, \phi'_n\}$ are spectral decompositions of D and D' respectively, then $\{\{\lambda_n\} \cup \{\lambda'_n\}, \{\phi_n \oplus 0\} \cup \{0 \oplus \phi'_n\}\}$ is a spectral decomposition of $D + D'$. Therefore

$f(t,x,D+D') = f(t,x,D) + f(t,x,D') \Rightarrow B_n(x,D+D') = B_n(x,D) + B_n(x,D')$.

As a corollary to (2) we obtain that the B_n are in general not trivial, for one can compute that $B_0(x,D) = (4\pi)^{-m/2}\dim V \neq 0$, so that $B_n(x,D+a) = \frac{(-a)^n}{n!} B_0(x,D) +$ higher order terms $\neq 0$ for $a > 0$.

The tensor product of two elliptic operators.

Let E and E' be Hermitian vector bundles over two different manifolds M and M' respectively, and

$$D: C^{\infty}(E) \to C^{\infty}(E), \qquad D': C^{\infty}(E') \to C^{\infty}(E')$$

positive, self-adjoint, elliptic operators .
Define $D \otimes D': C^{\infty}(E) \otimes C^{\infty}(E') \to C^{\infty}(E) \otimes C^{\infty}(E')$ by setting $D \otimes D' = D \otimes 1 + 1 \otimes D'$, where $C^{\infty}(E) \otimes C^{\infty}(E')$ is considered as a vector bundle over $M \times M'$.

For this definition to be of any use we need

$$D \otimes D': L^2(E \otimes E') \to L^2(E \otimes E')$$

and indeed this is so, since the Stone-Weierstrass theorem implies that any section $\theta(x,y)$ can be approximated in the L^2 norm by something of the form $\Sigma\theta_i(x)\cdot\theta_i'(y)$, which yields an isomorphism between $L^2(E) \otimes L^2(E')$ and $L^2(E \otimes E')$.

Proposition: D, D', $D \otimes D'$ as above. Then

$$B_n((x,x'), D \otimes D') = \sum_{p+q=n} B_p(x,D)B_q(x',D') \quad \text{(Künneth formula).}$$

Proof: Let $\{\lambda_i,\phi_i\}$ and $\{\lambda'_j,\phi'_j\}$ be spectral decompositions of D and D' respectively. One easily checks that $\{\lambda_i+\lambda'_j, \phi_i \otimes \phi'_j\}_{i,j}$ is a spectral decomposition of $D \otimes D'$. Therefore

$$f(t(x,x'), D \otimes D') = \sum_{i,j} e^{-\lambda_i t} e^{-\lambda'_j t} (\phi_i,\phi_i)(x)(\phi'_j,\phi'_j)(x')$$

$$= f(t,x,D)\, f(t,x',D')$$

$$\Rightarrow B_n((x,x'), D \otimes D') = \sum_{p+q=n} B_p(x,D)B_q(x',D').$$

This last proposition explains why we chose to work with the operator e^{-tD} instead of D^s. The trace of the kernel of the latter along the diagonal is $\sum_{n=0}^{\infty} \lambda_n^s(\phi_n,\phi_n)(x)$. For $D \otimes D'$ this equals

$$\sum_{i,j} (\lambda_i + \lambda'_j)^s(\phi_i,\phi_i)(x)(\phi'_j,\phi'_j)(x')$$

$$\neq \left(\sum_i \lambda_i^s(\phi_i,\phi_i)(x)\right)\left(\sum_j \lambda_j'^s(\phi'_j,\phi'_j)(x')\right) .$$

The previous proposition has the following generalization to elliptic complexes.

Let $(\mathcal{E},d) = \{E_i,d_i\}$ and $(\mathcal{E}',d') = \{E_j,d'_j\}$ be elliptic complexes over M and M' respectively. Define the elliptic complex $(\mathcal{E}'',d'') = (\mathcal{E} \otimes \mathcal{E}', d \otimes d')$ over $M \times M'$ by setting $E''_k = \bigoplus_{i+j=k} (E_i \otimes E'_j)$ and defining

$d''_k : C^\infty(E''_k) \to C^\infty(E''_{k+1})$ by $d''_k = \bigoplus_{i+j=k} (d_i \otimes 1 + (-1)^i 1 \otimes d'_j)$. The factor $(-1)^i$ ensures that $d''_{k+1} d''_k = 0$. Straightforward application of the two previous propositions gives us

Proposition: $B_n((x,x'), d \otimes d') = \sum_{p+q=n} B_p(x,d) B_q(x',d')$.

Proof: $$B_n((x,x'), d \otimes d') = \sum_{i,j=0}^{n} (-1)^{i+j} B_n((x,x'), d_i \otimes d'_j)$$

$$= \sum_{i,j=0}^{n} (-1)^{i+j} \sum_{p+q=n} B_p(x,d_i) B_q(x',d'_j)$$

$$= \sum_{p+q=n} \left[\left(\sum_{i=0}^{n} (-1)^i B_p(x,d_i) \right) \left(\sum_{j=0}^{n} (-1)^j B_q(x',d'_j) \right) \right]$$

Example: Let $(\Lambda(M),d)$ and $(\Lambda(M'),d')$ be the deRham complexes of M and M' respectively. Then

$$\Lambda(M \times M') = \Lambda(M) \otimes \Lambda(M') \quad \text{where}$$

$$\Lambda^k(M \times M') = \bigoplus_{i+j=k} (\Lambda^i(M) \otimes \Lambda^j(M'))$$

and $(d \otimes d')_k = \bigoplus_{i+j=k} (d_i \otimes 1 + (-1)^i 1 \otimes d'_j)$.

Hence for the de Rham complex of a product manifold we have

$$B_n((x,x'), d \otimes d') = \sum_{p+q=n} B_p(x,d)\, B_q(x',d').$$

The Index Theorem for the de Rham Complex.

We now know that for the de Rham complex of a manifold M of dimension m $\chi(M) = \text{index}(d) = \int_M B_m(x,d)\,dvol$.
We shall show that $B_m(x,d) = (2\pi)^{-m/2}$ (Pfaffian) which will yield a new proof of Chern's generalization of the Gauss-Bonnet theorem to even-dimensional manifolds. First we'll give another proof of the assertion $\chi(M) = 0$ for m odd.

Proposition: m odd $\Rightarrow B_m(x,d) = 0$.

Proof: We know that the Hodge $*$ operator commutes with all the Laplacians of the de Rham complex. If $\{\lambda_n^q, \phi_n^q\}$ is a spectral decomposition for

$D_q = d_{q-1} d^*_{q-1} + d^*_q d_q : \Lambda^q(M) \to \Lambda^q(M)$, then, since $*$ is an isometry and $*D_q = D_{m-q}*$, it follows that $\{\lambda_n^q, *\phi_n^q\}$ is a spectral decomposition for $D_{m-q}: \Lambda^{m-q}(M) \to \Lambda^{m-q}(M)$

$$\Rightarrow B_n(x,D_q) = B_n(x,D_{m-q}) \quad \text{for all } n \text{ and } q$$

$$\Rightarrow B_n(x,d) = \sum_{q=0}^{m} (-1)^q B_n(x,D_q) = 0 \quad \text{for all } n.$$

In particular $B_m(x,d) = 0 \Rightarrow \text{index }(d) = \chi(M) = 0$.

<u>Notation</u>: Let $\mathcal{P}$ be the ring of homogeneous polynomials in the indeterminants $g_{ij/\omega}$. If $A = g_{i_1 j_1/\omega_1} \cdots g_{i_r j_r/\omega_r}$ is a monomial of $P \in \mathcal{P}$, define

$$\text{ord } A = \text{order of } A = |\omega_1| + \ldots + |\omega_r|$$

$$L(A) = \text{length of } A = r$$

$\deg_k A$ = total number of times k appears as an index in A.

e.g. $A = g_{11/22}$ has ord $A = 2$, $L(A) = 1$, $\deg_1 A = \deg_2 A = 2$.

Given a local system consisting of a coordinate chart X about x_0 and a metric G, one can evaluate $P \in \mathcal{P}$ at x_0. Denote this value by $P(X,G)(x_0)$.

$P \in \mathcal{P}$ is said to be <u>invariant</u> if $P(X,G)(x_0) = P(Y,G)(x_0)$ for all charts X and Y about x_0, i.e. the value of P at x_0 depends only on G.

As in the 2-dimensional case we have

<u>Lemma</u>: $P \in \mathcal{P}$ and $P(X,G)(x_0) = 0$ for all $(X,G) \Rightarrow P \equiv 0$.

<u>Proof</u>: Similar to the 2-dimensional case.

We shall always assume that every chart X defines a normal coordinate neighborhood of x_0. Thus if $A = g_{i_1 j_1/\omega_1} \cdots g_{i_r j_r/\omega_r}$ is a monomial of $P \in \mathcal{P}$, $|\omega_k| \geq 2$ for $k = 1, \ldots r$. For $P \in \mathcal{P}$ we shall write $P = \Sigma c_P(A) A$ where $c_P(A)$ is the coefficient of the monomial A in P.

Note that m = dimension of M is fixed and $\mathcal{P}$ consists of polynomials in the indeterminates $g_{ij/\omega}$ which are to represent derivatives of the components of the metric tensor on m-manifolds. The next theorem is the key step in proving the index theorem for the de Rham complex.

<u>Theorem</u>: Let $\mathcal{R} \subseteq \mathcal{P}$ be the set of all invariant polynomials of order m which vanish identically on manifolds of the type $M^m = S^1 \times N^{m-1}$ where $\dim N = m-1$ and M is provided with the product metric (S^1 has the usual flat metric). Then

(1) $P \in \mathcal{R}$ and order $P = n < m \Rightarrow P \equiv 0$

(2) $\dim \{P \in \mathcal{R}\text{: order } P = m\} \leq 1$.

<u>Proof:</u> Since S^1 is flat $\mathcal{R}$ is precisely the ideal generated by the relation $g_{ij/\omega} = 0$ if $\deg_1 (g_{ij/\omega}) \neq 0$. Thus

$$P = \Sigma c_P(A)A \in \mathcal{R} \text{ iff } c_P(A) \neq 0 \Rightarrow \deg_1 A \neq 0.$$

Given a chart X, define the new chart Y by

$$y_k = x_{\tau(k)} \text{ where } \tau \text{ is a permutation of } \{1,\ldots,m\}.$$

Then $g^Y_{ij/\omega} = g^X{}_{\tau(i)\tau(j)/\tau(\omega)}$ so that $A^Y = A^X_\tau$ for monomials A. For $P \in \mathcal{R}$, $P(Y,G) = \Sigma c_P(A)A^Y = \Sigma c_P(A_\tau)A^X{}_\tau$

$$= P(X,G) = \Sigma c_P(A)A^X$$

$\Rightarrow c_P(A) = c_P(A_\tau)$. Therefore

(i)- $\deg_1 A \neq 0 \Rightarrow \deg_k A \neq 0$ for all A such that $C_P(A) \neq 0$ and all k

Now choose the chart Y such that $y_k = -x_k$, $y_i = x_i$ for $i \neq k$.

(ii)- Then $A^Y = (-1)^{\deg_k A} A$ for any monomial $A \Rightarrow$ every monomial A of $P \in \mathcal{R}$ has $\deg_k A$ even.

(iii)-(i) and (ii) imply that if A is a monomial of $P \in \mathcal{R}$ then $\deg_k A \geqslant 2$

Now let $A = g_{i_1 j_1/\omega_1} \cdots g_{i_r j_r/\omega_r}$ be any monomial of P. (iii) implies

$$2m \leqslant \sum_{k=1}^{m} \deg_k A = 2r + \text{ord } A.$$

But $\text{ord } A = \sum_{i=1}^{r} |\omega_i| \geqslant 2r$ (normal coordinates) and $\text{ord } A \leqslant m$ by assumption.

Therefore $2m \leqslant 2r + \text{ord } A \leqslant 2 \text{ ord } A \leqslant 2m$ so that $2m = 2r + \text{ord } A = 2\cdot\text{ord} A$ from which it follows that

(a) $2r = m = \text{ord } A$

(b) $|\omega_i| = 2$; $i = 1,\ldots,r$.

(c) $\deg_k A = 2$; $k = 1,\ldots,m$.

(1) Of the theorem now follows from (a).

To prove (2) of the theorem it is sufficient to show that if $P \in \mathcal{R}$ and order $P = m$ then P contains a monomial A of the form $A = g_{11/22} g_{33/44} \cdots g_{m-1,m-1/mm}$. For if $Q \in \mathcal{R}$, $P \neq Q$, choose λ so that $P - \lambda Q$ does not contain the monomial $g_{11/22} \cdots g_{m-1,m-1/mm}$. Since

$P - \lambda Q \in \mathcal{R}$, this implies $(P-\lambda Q)(X,G) = 0$ for all X,G, so that $P = \lambda Q$ by the lemma $\Rightarrow \dim \{P \in \mathcal{R}: \text{order } P = m\} \leqslant 1$.

First we show that some monomial of P is of the form $A = g_{11/\omega_1} \cdots$. Suppose not. Let A be any monomial of P, say $A = g_{ij/\omega_1} \cdots$. By permuting coordinates we can assume $i = 1$, $j = 2$, i.e. $A = g_{12/\omega_1} \cdots$. Let Y be the chart defined by the rotation

$$\frac{\partial}{\partial y_1} = a\frac{\partial}{\partial x_1} + b\frac{\partial}{\partial x_2}$$

$$\text{with } a^2 + b^2 = 1$$

$$\frac{\partial}{\partial y_2} = -b\frac{\partial}{\partial x_1} + a\frac{\partial}{\partial x_2}$$

$$\frac{\partial}{\partial y_k} = \frac{\partial}{\partial x_k} \quad ; \quad k > 2$$

P^Y can be computed by sending $1 \to a1 + b2$, $2 \to -b1 + a2$ formally.

Then $P^Y = f(a,b)A_0^X$ + other terms where $A_0^X = g_{11/\omega_1} \cdots$ and $f(a,b)$ is homogeneous of order 4 in (a,b). $f(a,b) = 0$ for all (a,b) s.t. $a^2 + b^2 = 1 \Rightarrow f \equiv 0$. Let B be a monomial of P such that $B^Y = \lambda_B a^3 b A_0 + \cdots = \lambda_B a^3 b g_{11/\omega_1} \cdots$ + other terms. λ_B is a constant. Then $B^X \to A_0^X$ by changing $1 \to 2$ or $2 \to 1$. If $B \to A_0$ by changing $1 \to 2$, then $\deg_2 B^X = \deg_2 A_0^X - 1 = 0$ which is impossible by (c) above. Therefore $B \to A_0$ by changing $2 \to 1$.

Now $A_0 = g_{11/\omega_1} \ldots 1 \ldots$ (this notation means 1 appears as an index in a term of A_0 other than g_{11/ω_1}). Therefore B is either $g_{11/\omega_1} \ldots 2 \ldots$ or $g_{12/\omega_1} \ldots 1 \ldots$. But $B = g_{11/\omega_1} \ldots$ is excluded by hypothesis. Therefore $B = g_{12/\omega_1} \ldots 1 \ldots = A$ is the only possibility so that $A^Y = B^Y = -a^3 b A_o^X + \ldots$.

This contradicts the fact that $f(a,b) \equiv 0$.

Thus we can choose a monomial A in P such that $A = g_{11/\omega_1} A'$ such that 1 does not appear as an index in ω_1 or A', since $\deg_1 A = 2$ by (c).

Now we claim it is possible to choose A such that $A = g_{11/22} \ldots$. Suppose not. Without loss of generality we can assume $A = g_{11/23} \ldots$. Using the same argument as before with $A_0 = g_{11/22} \ldots$ we can arrive at a contradiction.

Proceeding by induction we obtain the existence of

$A = g_{11/22} \ldots g_{m-1,m-1/mm}$ in P.

Let $\Omega = (\Omega_{ij})_{i,j=1}^m$ be the curvature matrix of 2-forms associated to the Levi-Civita connection on TM, m = dimension M, m even. Define Pffafian $(\Omega) = \mathrm{Pff}(\Omega) = * \left(\Sigma \varepsilon(i_1 \ldots i_m) \Omega_{i_1 i_2} \wedge \ldots \wedge \Omega_{i_{m-1}, i_m}\right)$. The summation is over all permutations $(i_1 \ldots i_m)$ of $\{i, \ldots, m\}$ and $\varepsilon(i_1 \ldots i_m)$ is the sign of the permutation. One easily checks that $\mathrm{Pff}(\Omega)$ has the following properties:

(1) $\mathrm{Pff}(\Omega)$ is defined invariantly with respect to change of local frame for TM.

(2) $\mathrm{Pff}(\Omega)$ does not depend on the orientation of any local frame for TM

(3) $\mathrm{Pff}(\Omega)$ is a homogeneous polynomial of degree m in the derivatives of the metric which is not identically zero.

(4) $\mathrm{Pff}(\Omega) = 0$ on manifolds of the type $S^1 \times N^{m-1}$.

Theorem. (Index theorem for the de Rham complex.)

(i) $B_n(x,d) = 0$ for $n < m$ and n odd

(ii) For m even, index $(d) = \chi(M) = \left(\frac{1}{2\pi}\right)^{m/2} \int_M \mathrm{Pffafian}\ (\Omega) \mathrm{dvol}$.

Proof: For M of the type $M = S^1 \times N^{m-1}$ and $x = (x_1, \ldots, x_m) \in M$ we have by an earlier result that

$$B_n(x,d_M) = \sum_{p+q=n} B_p(x_1,d_{S^1}) B_q(x_2,\ldots,x_m),d_N), \quad \text{for all } n.$$

But $B_p(x_1,d_{S^1}) = 0$ for all $p \Rightarrow B_n(x,d_M) = 0$ on all manifolds of the type $M = S^1 \times N^{m-1}$ and all n. (i) now follows immediately from the previous theorem. Since in particular $B_m(x,d) = 0$ on manifolds of the type $S^1 \times N^{m-1}$, we know by the previous theorem and (4) above that $B_m(x,d) = \lambda_m \mathrm{Pff}(\Omega)$. To evaluate λ_m we use the fact that the Pffafian is multiplicative on product manifolds.

Thus on $\underbrace{S^2 \times \ldots \times S^2}_{m/2 \text{ factors}}$

$$\mathrm{Pff}\left(\Omega_{S^2\times\ldots\times S^2}\right) = \left[\mathrm{Pff}\left(\Omega_{S^2}\right)\right]^{m/2} \quad \text{so that}$$

$$2^{m/2} = \chi(S^2\times\ldots\times S^2) = \int_{S^2\times\ldots\times S^2} B_m(x,d)\,dvol = \lambda_m \int_{S^2\times\ldots\times S^2} \mathrm{Pff}\left(\Omega_{S^2\times\ldots\times S^2}\right) dvol$$

$$= \lambda_m \int_{S^2} \mathrm{Pff}\left(\Omega_{S^2}\right) dvol^{\ m/2} = \lambda_m (4\pi)^{m/2}.$$

Therefore $\lambda_m = \left(\frac{1}{2\pi}\right)^{m/2}$, which proves (ii).

Chapter V

The Rieman-Roch-Hirzebruch Formula for the Arithmetic Genus

Let M be a compact 2m-dimensional real manifold. M is said to be a complex analytic manifold of complex dimension m if there are coordinate charts

$$\theta_\alpha : U_\alpha \to \{z \in \mathbb{C}^m : \|z\| < 1\} \text{ such that}$$

$$\theta_\alpha \circ \theta_\beta^{-1} : \theta_\beta(U_\alpha \cap U_\beta) \to \theta_\alpha(U_\alpha \cap U_\beta) \text{ is holomorphic.}$$

If $z_a = x_a + iy_a$, $a = 1,\ldots,m$, are complex local coordinates for M then

$$\left.\begin{aligned} dz_a &= dx_a + idy_a \\ d\bar{z}_a &= dx_a - idy_a \end{aligned}\right\} \text{ form a basis for } T^*M \otimes \mathbb{C}$$

and

$$\left.\begin{aligned} \frac{\partial}{\partial z_a} &= \frac{1}{2}\left(\frac{\partial}{\partial x_a} - i\,\frac{\partial}{\partial y_a}\right) \\ \frac{\partial}{\partial \bar{z}_a} &= \frac{1}{2}\left(\frac{\partial}{\partial x_a} + i\,\frac{\partial}{\partial y_a}\right) \end{aligned}\right\} \text{ form a basis for } TM \otimes \mathbb{C}$$

(TM and T^*M are the real tangent and cotangent bundles of M). There is a well-defined involution: $J: TM \to TM$ which is given in local coordinates by

$$J\left(\frac{\partial}{\partial x_a}\right) = \frac{\partial}{\partial y_a} \quad , \quad J\left(\frac{\partial}{\partial y_a}\right) = -\frac{\partial}{\partial x_a} \quad .$$

J can be extended to $J\colon TM \otimes C \to TM \otimes C$ in the obvious way: $J(X + iY) = JX + iJY$ for $X, Y \in TM$.

The Dolbeault Complex

The exterior algebra of the complexified cotangent bundle is

$$\Lambda(T^*M \otimes C) = \bigoplus_{n=0}^{2m} \Lambda^n(T^* \otimes C)$$

$$\Lambda^n(T^*M \otimes C) = \bigoplus_{p+q=n} \Lambda^{p,q}(T^*M \otimes C) \quad \text{where}$$

$$\Lambda^{p,q}(T^*M \otimes C) = \text{span} \ \{dz_{i_1} \wedge \cdots \wedge dz_{i_p} \wedge d\bar{z}_{j_1} \wedge \cdots \wedge d\bar{z}_{j_q} \quad \text{with}$$

$$i_1 < \cdots < i_p \quad \text{and} \quad j_1 < \cdots < j_q\}$$

We define the operators ∂ and $\bar{\partial}$ as follows:

For $f \in C^\infty(\Lambda^{0,0})$ define

$$\partial f = \sum_{a=1}^{m} \frac{\partial f}{\partial z_a} dz_a \in C^\infty(\Lambda^{1,0}) \quad \text{and}$$

$$\bar{\partial} f = \sum_{a=1}^{m} \frac{\partial f}{\partial \bar{z}_a} d\bar{z}_a \in C^\infty(\Lambda^{0,1}).$$

Then extend ∂: $C^\infty(\Lambda^{p,q}) \to C^\infty(\Lambda^{p+1,q})$ by defining

$$\partial(f dz_{i_1} \wedge \cdots \wedge dz_{i_p} \wedge d\bar{z}_{j_1} \wedge \cdots \wedge d\bar{z}_{j_q}) = \partial f \cdot dz_{i_1} \wedge \cdots \wedge d\bar{z}_{j_q}$$

and extend by linearity to all of $C^\infty(\Lambda^{p,q})$. Similarly $\bar\partial$ can be extended so that $\bar\partial : C^\infty(\Lambda^{p,q}) \to C^\infty(\Lambda^{p,q+1})$. Observe that $d = \partial + \bar\partial$, for if $f \in C^\infty(\Lambda^{Q,0}) = C^\infty(\Lambda^0)$, then

$$(\partial + \bar\partial)f = \sum_{a=1}^{m} \left(\frac{\partial f}{\partial z_a}\, dz_a\right) + \left(\frac{\partial f}{\bar z_a}\, d\bar z_a\right)$$

$$= \sum_{a=1}^{m} \frac{1}{2}\left(\frac{\partial f}{\partial x_a} - i\,\frac{\partial f}{\partial y_a}\right)(dx_a + i dy_a) + \frac{1}{2}\left(\frac{\partial f}{\partial x_a} + i\frac{\partial f}{\partial y_a}\right)(dx_a - i dy_a)$$

$$= \sum_{a=1}^{m} \left(\frac{\partial f}{\partial x_a}\, dx_a + \frac{\partial f}{\partial y_a}\, dy_a\right)$$

The complex analogue of the de Rham complex is the Dolbeault complex

$$0 \to C^\infty(\Lambda^{0,0}) \xrightarrow{\bar\partial} C^\infty(\Lambda^{0,1}) \xrightarrow{\bar\partial} \cdots\cdots \xrightarrow{\bar\partial} C^\infty(\Lambda^{0,m}) \to 0.$$

One checks easily that

(1) $\bar\partial^2 = 0$.

(2) $\sigma(\bar\partial)$ = exterior multiplication by $\bar\xi$.

(3) the Dolbeault complex is exact at the symbol level (proofs very similar to the case of the de Rham complex). Thus the Dolbeault complex is an elliptic complex. It turns out that for the special class of Kaehler manifolds the index theorem for the Dolbeault complex is the Riemann-Roch-Hirzebruch theorem. We shall be giving an analytic proof of this theorem

Now $\bar{\partial} f = 0 \Leftrightarrow f$ satisfies the Cauchy-Riemann equations

$\Leftrightarrow f$ is holomorphic

Thus if E is a holomorphic vector bundle over M we can form the <u>twisted Dolbeault complex</u>, which will be invariantly defined;

$$0 \to C^\infty(\Lambda^{0,0} \otimes E) \xrightarrow{\bar{\partial}} C^\infty(\Lambda^{0,1} \otimes E) \xrightarrow{\bar{\partial}} \cdots \xrightarrow{\bar{\partial}} C^\infty(\Lambda^{0,m} \otimes E) \to 0$$

The results for the Dolbeault complex extend naturally to the twisted Dolbeault complex. For Kaehler manifolds,

<u>Theorem:</u>

(1) $B_n(x,\bar{\partial}) = 0$ for $n < 2m$.

(2) $B_{2m}(x,\bar{\partial})$ is a polynomial in the characteristic classes of M and E. We will only prove this if $E = 1$; from the fact that

$$\text{arith. genus } (M) = \text{index } (\bar{\partial}) = \int_M B_{2m}(x,\bar{\partial})$$

and the fact that B_{2m} is a polynomial in the characteristic classes at M, the Riemann-Roch theorem follows easily.

Kaehler manifolds.

Given a compact complex manifold M with a Riemannian metric G on TM, G can be extended to be bilinear over $TM \otimes \mathbb{C}$. If G also satisfies $G(X,Y) = G(JX,JY)$ for all X,Y in $TM \otimes \mathbb{C}$, then (M,G) is called a Riemannian complex manifold. Such metrics can always be constructed using a partition of unity. Since $J^2 = -1$, it follows that

$$G\left(\frac{\partial}{\partial z_a}, \frac{\partial}{\partial z_b}\right) = 0 = G\left(\frac{\partial}{\partial \bar{z}_a}, \frac{\partial}{\partial \bar{z}_b}\right) \qquad \text{for such metrics.}$$

Define

$$\Omega(X,Y) = G(\mathbf{X},Y) \qquad \text{for} \quad X,Y \in T\,M \otimes C$$

Then

$$\Omega(X,Y) = G(JX,Y) = G(Y,JX)$$

$$= G(JY,J^2X) = G(JY,-X)$$

$$= -G(JY,X) = -\,\Omega(Y,X).$$

Thus Ω defines a 2-form on M [illegible], called the fundamental 2-form. (M,G) is said to be Kaehler if $d\Omega = 0$.

Note that since

$$\Omega\left(\frac{\partial}{\partial z_a},\frac{\partial}{\partial \bar{z}_b}\right) = G\left(J\frac{\partial}{\partial z_a},\frac{\partial}{\partial \bar{z}_b}\right) = G\left(-i\frac{\partial}{\partial z_a},\frac{\partial}{\partial \bar{z}_b}\right) = -i\; G\left(\frac{\partial}{\partial z_a},\frac{\partial}{\partial \bar{z}_b}\right)$$

it follows that $\Omega = \frac{i}{2} \sum_{a,b=1}^{m} g_{a\bar{b}}\, dz_a \wedge d\bar{z}_b$ where $g_{a\bar{b}} = G\left(\frac{\partial}{\partial z_a}, \frac{\partial}{\partial \bar{z}_b}\right)$.

We will now exploit the Kaehler hypothesis to construct normalized holomorphic coordinates. We first study a complex Kaehler surface.

Normal coordinates on Kaehler manifolds

Suppose M^2 is a Kaehler manifold, $z_0 \in M^2$ and (Z_1, Z_2) are local holomorphic coordinates centered at z_0. Let $g_{a\bar{b}}$ (which we can assume $= \delta_{ab} + O(Z))$ be the components of the metric tensor in this coordinate system. We want to find new coordinates $(\tilde{Z}_1, \tilde{Z}_2)$ such that the components of the corresponding metric tensor $\tilde{g}_{a\bar{b}}$ have the property $\tilde{g}_{a\bar{b}/c}(z_0) = 0$ for all $a, b, c = 1, 2$.

Let $\tilde{Z}_1, \tilde{Z}_2$ be given by

$$Z_1 = \tilde{Z}_1 + \frac{1}{2} a^1_{11}\tilde{Z}_1^2 + a^1_{12}\tilde{Z}_1\tilde{Z}_2 + \frac{1}{2} a^1_{22}\tilde{Z}_2^2 + O(\tilde{Z}^3)$$

$$Z_2 = \tilde{Z}_2 + \frac{1}{2} a^2_{11}\tilde{Z}_1^2 + a^2_{12}\tilde{Z}_1\tilde{Z}_2 + \frac{1}{2} a^2_{22}\tilde{Z}_2^2 + O(\tilde{Z}^3)$$

where the a^i_{jk} are constants to be determined.

Using the chain rule $\dfrac{\partial}{\partial\tilde{Z}_a} = \sum\limits_{b=1}^{2} \dfrac{\partial Z_b}{\partial\tilde{Z}_a}\dfrac{\partial}{\partial Z_b} + \sum\limits_{b=1}^{2} \dfrac{\partial\bar{Z}_b}{\partial\bar{\tilde{Z}}_a}\dfrac{\partial}{\partial\bar{Z}_b}$ we get

$$\frac{\partial}{\partial\tilde{Z}_1} = (1 + a^1_{11}\tilde{Z}_1 + a^1_{12}\tilde{Z}_2)\frac{\partial}{\partial Z_1} + (a^2_{11}\tilde{Z}_1 + a^2_{12}\tilde{Z}_2)\frac{\partial}{\partial Z_2} + \cdots$$

$$\frac{\partial}{\partial\tilde{Z}_2} = (a^1_{12}\tilde{Z}_1 + a^1_{22}\tilde{Z}_2)\frac{\partial}{\partial Z_1} + (1 + a^2_{12}\tilde{Z}_1 + a^2_{22}\tilde{Z}_2)\frac{\partial}{\partial Z_2} + \cdots$$

from which we can compute

$$\tilde{g}_{2\bar{2}} = g_{2\bar{2}} + a^2_{12}\tilde{Z}_1 + a^2_{22}\tilde{Z}_2 + O(\tilde{Z}^2) + \text{ terms in } \bar{\tilde{Z}}$$

$$\tilde{g}_{1\bar{2}} = g_{1\bar{2}} + a^2_{11}\tilde{Z}_1 + a^2_{12}\tilde{Z}_2 + O(\tilde{Z}^2) + \text{ terms in } \bar{\tilde{Z}}\ .$$

Therefore

$$\left.\begin{aligned}\tilde{g}_{2\bar{2}/1} &= g_{2\bar{2}/1} + a^2_{12} \\ \tilde{g}_{1\bar{2}/2} &= g_{1\bar{2}/2} + a^2_{12}\end{aligned}\right\} \circledast$$

The fundamental 2-form $\Omega = \frac{i}{2}\{g_{1\bar{1}}dz_1 \wedge d\bar{z}_1 + g_{1\bar{2}}dz_1 \wedge d\bar{z}_2 + g_{2\bar{1}}\, dz_2 \wedge d\bar{z}_1$

$$+ g_{2\bar{2}}dz_2 \wedge d\bar{z}_2\} .$$

M^2 Kaehler $\Leftrightarrow d\Omega = 0$

$\Leftrightarrow g_{1\bar{1}/2}\, dz_2 \wedge dz_1 \wedge d\bar{z}_1 + g_{1\bar{2}/2} dz_2 \wedge dz_1 \wedge d\bar{z}_2 + g_{2\bar{1}/1}\, dz_1 \wedge dz_2 \wedge d\bar{z}_1 + g_{2\bar{2}/1}\, dz_1 \wedge dz_2 \wedge d\bar{z}_2$

$= \bar{\partial}(\Omega) = 0$

$\Leftrightarrow (-g_{1\bar{1}/2} + g_{2\bar{1}/1})dz_1 \wedge dz_2 \wedge d\bar{z}_2 + (g_{1\bar{2}/2} - g_{2\bar{2}/1})dZ_2 \wedge dZ_1 \wedge d\bar{Z}_2 = 0$

$\Leftrightarrow g_{1\bar{1}/2} = g_{2\bar{1}/1}$ and $g_{1\bar{2}/2} = g_{2\bar{2}/1}$ (**)

In $\circledast$ above set $a^2_{12} = -g_{2\bar{2}/1}(Z_0)$.

Then $\tilde{g}_{2\bar{2}/1}(z_0) = 0$, and by (**) and $\circledast$

$\tilde{g}_{1\bar{2}/2}(z_0) = 0$.

Similarly we can get $\tilde{g}_{2\bar{2}/2}$, $\tilde{g}_{1\bar{2}/1}$, $\tilde{g}_{1\bar{1}/1}$, $\tilde{g}_{1\bar{1}/2}$, $\tilde{g}_{2\bar{1}/1}$, $\tilde{g}_{2\bar{1}/2}$ to vanish at z_0 . It is precisely the Kaehler conditions (**) which enable us to determine 6 constants a^i_{jk} which make 8 quantities $\tilde{g}_{a\bar{b}/c}(Z_0)$ vanish.

Using

$$g_{b\bar{a}/c} = g_{c\bar{a}/b} \Leftrightarrow \overline{g_{b\bar{a}/c}} = \overline{g_{c\bar{a}/b}} \Leftrightarrow g_{a\bar{b}/c} = g_{a\bar{c}/\bar{b}}$$

we can also ensure that $g_{a\bar{b}/\bar{c}}(z_0) = 0$.

The following theorem generalizes what we have just done on 2-manifolds to higher dimensions and also asserts that we can choose a coordinate system at any fixed point so that as many holomorphic derivatives of the metric as we please vanish at that point. Such a coordinate system is called normal.

Theorem: M is Kaehler $\Leftrightarrow$ for each $z_0 \in M$ there is a holomorphic normal coordinate system centered at z_0 .

Proof: One direction is easy. Since $d\Omega$ involves only first order derivatives of the metric, the existence of normal coordinates implies $d\Omega = 0$, since the latter condition is independent of the coordinate system.

Now suppose M^m is Kaehler with fundamental 2-form Ω . Ω is a 1-1 form $\Rightarrow d\Omega = \partial\Omega + \bar{\partial}\Omega$ with $\partial\Omega$ a (2,1) form and $\bar{\partial}\Omega$ a (1,2) form.

Thus $d\Omega = 0 \Leftrightarrow \partial\Omega = 0$ & $\bar{\partial}\Omega = 0$.

$$\Leftrightarrow g_{a\bar{b}/c} = g_{c\bar{b}/a}$$

$$\Leftrightarrow g_{a\bar{b}/\bar{c}} = g_{a\bar{c}/\bar{b}}$$

$\omega = (\omega(1), \ldots, \omega(m))$ will denote a multi-index such that $|\omega| = 2$. Let $(Z_1,\ldots,Z_m)$ be coordinates about z_0. Choose new coordinates $(\tilde{Z}_1,\ldots,\tilde{Z}_m)$ about z_0 such that

$$Z_a = \tilde{Z}_a + \sum_{|\omega|=2} c_{a,\omega} \tilde{Z}^{\omega} + \text{higher powers of } \tilde{Z} , \quad a = 1, \ldots, m$$

where the $c_{a,\omega}$ are constants to be determined.

Define

$$\left.\begin{aligned}\omega_b(b) &= \omega(b) - 1\\ \omega_b(c) &= \omega(c)\ ,\quad \text{if}\quad \omega(c) \neq 0\\ \omega_b(c) &= \quad \emptyset\ ,\quad \text{if}\quad \omega(c) = 0\end{aligned}\right\}\quad \text{for}\quad b,\ c = 1,\ \ldots,\ m\ .$$

Then

$$\frac{\partial}{\partial \tilde{Z}_a} = \frac{\partial}{\partial Z_a} + \left(\sum_{|\omega|=2} c_{b,\omega}\ \omega(a)\tilde{Z}^{\omega_a}\right)\frac{\partial}{\partial Z_b}\ + \text{other terms}$$

$$\frac{\partial}{\partial \tilde{Z}_b} = \frac{\partial}{\partial Z_b} + \left(\sum_{|\omega|=2} c_{a,\omega}\ \omega(b)\tilde{Z}^{\omega_b}\right)\frac{\partial}{\partial Z_a}\ + \text{other terms}$$

from which we can compute

$$\tilde{g}_{a\bar{b}} = g_{a\bar{b}} + \sum_{\omega} c_{b,\omega}\ \omega(a)\tilde{Z}^{\omega_a} + O(\tilde{Z}^2) + \text{terms in}\ \bar{\tilde{Z}}\ .$$

Therefore if $c \neq a$,

$$\tilde{g}_{a\bar{b}/c} = g_{a\bar{b}/c} + c_{b,\omega'}$$

where ω' is that multi-index for which $\omega'(a) = 1 = \omega'(c)$ and all other components of ω' are zero. Thus

$$\tilde{g}_{a\bar{b}/c} = g_{a\bar{b}/c} + c_{a,bc} \quad \text{where} \quad c_{a,bc} = c_{b,\omega'}$$

and

$$\tilde{g}_{c\bar{b}/a} = g_{c\bar{b}/a} + c_{a,bc} \quad \text{since} \quad c_{c,ba} = c_{b,\omega'} = c_{a,bc} \, .$$

Set

$$c_{a,bc} = -g_{a\bar{b}/c}(z_0) \, .$$

Then

$$\tilde{g}_{a\bar{b}/c}(z_0) = 0$$

and by the Kaehler conditions also

$$\tilde{g}_{c\bar{b}/a}(z_0) = 0$$

As in the case of 2-manifolds, one can show that all $\tilde{g}_{a\bar{b}/c}$ and $\tilde{g}_{a\bar{b}/\bar{c}}$ vanish at z_0.

For $c = a$ we get $\tilde{g}_{a\bar{b}/a} = g_{a\bar{b}/a} + 2c_{b,\omega'}$ where ω' is that multi-index for which $\omega'(a) = 2$ and all other components of ω' are zero.

Setting $c_{b,\omega'} = c_{a,ba} = -\dfrac{g_{a\bar{b}/a}}{2}(z_0)$ we get $\tilde{g}_{a\bar{b}/a}(z_0) = 0$.

Similarly $\tilde{g}_{a\bar{b}/\bar{a}}(z_0) = 0$.

Next we prove by induction that there are holomorphic coordinates

$\widetilde{Z}_a$ near z_0 such that

$$\frac{\partial}{\partial z_{c_1}} \cdots \frac{\partial}{\partial z_{c_n}}(g_{a\bar{b}}) = g_{a\bar{b}/c_1 \ldots c_j}(z_0) = 0 \qquad 1 \leqslant j \leqslant m$$

$$\frac{\partial}{\partial \bar{z}_{c_1}} \cdots \frac{\partial}{\partial z_{c_j}} (g_{a\bar{b}}) = g_{a\bar{b}/\bar{c}_1 \ldots \bar{c}_j}(z_0) = 0 \qquad 1 \leqslant j \leqslant m$$

We have just proved it for $j = 1$. Assume that Z_a is a coordinate system such that $g_{a\bar{b}/(c_1,\ldots,c_j)}(z_0) = 0$ for $1 \leqslant j \leqslant n - 1$. Let the new coordinates $\widetilde{Z}_a$ be defined by

$$Z_a = \widetilde{Z}_a + \sum_{|\omega|=n+1} c_{a,\omega} \widetilde{Z}^{\omega}$$

where $c_{a,\omega}$ are constants to be determined.

Since $Z_a = \widetilde{Z}_a + 0(\widetilde{Z}^{n+1})$ it follows that $\widetilde{g}_{a\bar{b}/(c_1,\ldots,c_j)}$ $= g_{a\bar{b}/(c_1,\ldots,c_j)}(z_0) = 0$ for $1 \leqslant j \leqslant n - 1$. By an argument similar to the case $n = 1$, one gets for suitable ω'

(i) $$\widetilde{g}_{a\bar{b}/(c_1,\ldots,c_n)} = g_{a\bar{b}/(c_1,\ldots,c_n)} + \omega' ! c_{b,\omega'}$$

Differentiating the Kaehler condition gives

(ii) $$g_{a\bar{b}/(c_1,\ldots,c_n)} = g_{c_1\bar{b}/(a,c_2,\ldots,c_n)}$$

If we set $c_{b,\omega'} = c_{a}, bc_1 \ldots c_n = \dfrac{-g_{a\bar{b}/(c_1,\ldots,c_n)}}{\omega' !}(Z_0)$

then (i) and (ii) give

$$\tilde{g}_{a\bar{b}/(c_1,\ldots,c_n)}(z_0) = 0 \quad \text{for all} \quad a,\, b,\, (c_1,\ldots,c_n).$$

Similarly we can get $\tilde{g}_{a\bar{b}/(\bar{c}_1,\ldots,\bar{c}_n)}(z_0) = 0$.

The theorem just proved enables us to normalize the coordinate system and frame at z_0 uniquely up to unitary transformations of the frame and holomorphic coordinate transformations of arbitrarily high order.

The Index theorem for the Dolbeault complex of Kaehler manifolds (The Riemann-Roch theorem)

On a Kaehler manifold M , with metric tensor components $g_{a\bar{b}}$ define

$g(a_1a_2 \ldots ; b_1b_2 \ldots) = g_{a_1\bar{b}_1/a_2\ldots\bar{b}_2} \ldots$. Then

$$
\begin{aligned}
g(a_1a_2 \ldots ; b_1b_2 \ldots) &= g_{a_1\bar{b}_1/a_2\ldots\bar{b}_2} \ldots \\
&= g_{a_2\bar{b}_1/a_1\ldots\bar{b}_2} \ldots \\
&= g(a_2a_1\ldots;b_1b_2\ldots) .
\end{aligned}
$$

by the differentiated Kaehler condition.

Similarly, $g(a_1a_2 \ldots ; b_1b_2 \ldots) = g(a_1a_2 \ldots ; b_2b_1 \ldots)$. It follows that $g(a_1 \ldots ; b_1 \ldots)$ is symmetric in the a_i's and the b_j's , so we can write $g(\omega, \bar{\omega}')$ with $\omega = a_1a_2 \ldots$. $\omega' = b_1b_2\ldots$ instead of $g_{a_1\bar{b}_1/a_2\ldots\bar{b}_2\ldots}$. Thus $g(\omega, \bar{\omega}')$ is symmetric in the arguments of ω and ω' .

For Kaehler manifolds the analògue of the index theorem for the de Rham complex, is the index theorem for the Dolbeault complex, i.e. the Riemann-Roch theorem.

The next result is an essential step in an analytic proof of the Riemann-Roch theorem.

<u>Theorem</u>: Let M be a Kaehler manifold of complex dimension m with metric tensor components $g_{a\bar{b}}$. Let P be a homogeneous polynomial in the variables $g(\omega, \bar{\omega}')$ such that

(1) $P(T^2 \times N^{2m-2}) \equiv 0$ where T^2 is the flat 2-torus.

(2) P is invariant.

(3) The order of P in the derivatives of $g_{a\bar{b}} = n \leqslant 2m$.

$$\text{Then} \begin{cases} n < 2m \Rightarrow P \equiv 0 \\ \\ n = 2m \Rightarrow P \text{ is a combination of Chern classes of } TM . \end{cases}$$

Proof: (Outline.) The idea is similar to the proof of the corresponding theorem we obtained to prove the index theorem for the De Rham complex.

Let $A = g(\omega_1; \bar{\omega}'_1) \ldots g(\omega_k; \bar{\omega}'_k)$ be any monomial in the derivatives of the metric tensor. Define

$$\deg_a A = \omega_1(a) + \ldots + \omega_k(a)$$

$$\deg_{\bar{a}} A = \omega'_1(a) + \ldots + \omega'_k(a)$$

$$\text{Ord } A = |\omega_1| + \ldots + |\omega_k| + |\omega'_1| + \ldots + |\omega'_k| - 2k$$

(thus ord A = the total number of derivatives of the metric tensor in A)

$$L(A) = \text{length } A = k$$

$$C_P(A) = \text{the coefficient of } A \text{ in the polynomial } P .$$

We can suppose the coordinates Z are normalized at x_0 such that $g(a,b)(z_0) = \delta_{ab}$ and $g(\omega,\bar{\omega}')(z_0) = 0$ for $|\omega| = 1$ or $|\omega'| = 1$ and $|\omega| + |\omega'| > 2$.

We are permitted unitary changes of coordinates, which leave the coordinate system normalized, i.e. coordinate changes of the form

$$\frac{\partial}{\partial z'_a} = r\frac{\partial}{\partial z_a} + s\frac{\partial}{\partial z_b} \qquad \text{with } |r|^2 + |s|^2 = 1$$

$$\frac{\partial}{\partial z'_b} = -\bar{s}\frac{\partial}{\partial z_a} + \bar{r}\frac{\partial}{\partial z_b}$$

$$\frac{\partial}{\partial z'_c} = \frac{\partial}{\partial z_c} \quad \text{for } c \neq a, b .$$

Suppose A is a monomial of P and in the new coordinates Z' .

$A' = f(r,\bar{r}, s, \bar{s})A$ + other terms where $f(r, \bar{r}, s, \bar{s})$ is homogeneous.

As in the real case, if $f(r, \bar{r}, s, \bar{s}) = 0$ for all r, s s.t.

$|r|^2 + |s|^2 = 1$ then $f \equiv 0$.

Using such changes of coordinates one can prove

(a) If A is any monomial of P , P as in the statement of the theorem,

then $\deg_a A = \deg_{\bar{a}} A \neq 0$ for $a = 1, \ldots, m$.

(b) There is a monomial A of P of the form

$$A = g(11; \bar{\omega}_1) \ldots g(kk; \bar{\omega}_k)$$

(c) $|\omega_i| \geq 2$ for $i = 1, \ldots, k$ (normalized coordinates)

hence $\operatorname{ord} P = \operatorname{ord} A \geq 2k$

(d) $\deg_a A = 0$ for $a > k$, hence by (a) $k \geq m$.

Thus $2m \geq n = \operatorname{ord} P \geq 2k \geq 2m$.

$\Rightarrow n = 2m$

$\Rightarrow P \equiv 0$ if $\operatorname{ord} P < 2m$.

Moreover, when $n = 2m$ there is a monomial A of P of the form

$$A = g(11; \bar{\omega}_1) \ldots g(mm; \bar{\omega}_m) \quad \text{with } |\omega_i| = 2 \text{ for } i = 1, \ldots, m .$$

In fact one can show that for $n = 2m$ there must exist a monomial A of P of the form $A_\sigma = g(11; i_1 i_1) \ldots g(mm; i_m i_m)$, $\deg_{\bar{a}} A_\sigma \neq 0$ for $a = 1, \ldots, m$, where $\sigma(j) = i_j$ is a permutation of $\{1, \ldots, m\}$.

Let $L_\sigma P = C_{A_\sigma}(P)$. Then $P \neq 0 \Rightarrow L_\sigma P \neq 0$ for some $\sigma \in \mathrm{Perm}(m)$. In the case of the de Rham complex we had one classifying monomial, viz., $g_{11/22} \ldots g_{m-1,m-1/mm}$. Here we have more than one. The question is, how many?

If σ and τ belong to the same conjugacy class of $\mathrm{Perm}(m)$ they are essentially the same up to renumbering of indices. Thus if A_σ goes to A_τ under a renumbering of indices then $C_{A_\sigma}(P) = C_{A_\tau}(P)$. Therefore τ is conjugate to $\sigma \Leftrightarrow L_\tau = L_\sigma$. Now the number of distinct conjugacy classes of $\mathrm{Perm}(m) = \pi(m) =$ the number of distinct partitions of m . Hence $\{\dim$ invariant polynomials of order $2m$ which vanish on $T^2 \times N^{2m-2}\} \leq \pi(m)$ since there are at most $\pi(m)$ separating linear functionals L_σ on the space of such invariant polynomials.

Next we note that

dim{homogeneous polynomials of order m in the Chern classes of TM (order $2m$ in the derivatives of the metric} $= \pi(m)$

and that the Chern classes are invariant and vanish on manifolds of the type $T^2 \times N^{2m-2}$.

Therefore the dimension of the invariant polynomials under consideration precisely equals $\pi(m)$. Thus for $n = 2m$, $P =$ a homogeneous polynomial of order m in the Chern classes of TM .

Corollary: For the Dolbeault complex of a Kaehler manifold M

$B_n(x, \bar{\partial}) = 0$, $n < 2m$

$$\text{index}(\bar{\partial}) = \int_M B_{2m}(x, \bar{\partial})d\,\text{vol} = \int_M (\text{Riemann-Roch invariant})d\,\text{vol}$$

Proof: By the multiplicative property of the Dolbeault complex, $B_n(x, \bar{\partial})$ satisfy the properties of the polynomials P of the previous theorem. We compute $B_{2m}(x, \bar{\partial})$ on specific examples to show it is the Riemann-Roch invariant since it is a characteristic class.

Remarks. (1) A similar theorem is true for the twisted Dolbeault complex of a Kaehler manifold $\bar{\partial}: E \otimes \Lambda^{0,p} \to E \otimes \Lambda^{0,p+1}$ where E is a holomorphic vector bundle over M .

(2) In the generic non-Kaehler case it can be shown that

$B_n(x, \bar{\partial}) \neq 0$ for $n \geqslant m$, n-even

$B_{2m}(x, \bar{\partial}) \neq$ R.R. invariant

$B_n(x, \bar{\partial}) = 0$ for $n < m$.

However, on $S^1 \times S^3$, which is not Kaehler

$B_{2m}(x, \bar{\partial}) =$ constant $\times$ R.R. invariant

$B_n(x, \bar{\partial}) = 0$ for $n \neq 2m$.

Application. $\mathbb{C}P_2(\text{-orn})$ is not a complex manifold. Let $\mathbb{C}P_2(\text{-orn})$ denote complex projective space of dimension 2 with the opposite of the normal orientation.

For any manifold M^2 of complex dimension 2 it can be shown that $\operatorname{index}(\bar{\partial}) = \int_{M^2} (ac_1^2 + bc_2)\,d\,\mathrm{vol}$ where c_1 and c_2 are the Chern classes of M^2.

Thus $\operatorname{index}(\bar{\partial}) = \int_{M^2} (\alpha E + \beta p_1)\,d\,\mathrm{vol} = \alpha\chi(M) + \beta\,\operatorname{sign}(M)$

since the Euler class E and the first Pontryagin class p_1 are a basis for the characteristic classes of TM^2.

For $M^2 = S^2 \times S^2$, $\chi(M) = 4$, $\operatorname{sign}(M) = 0$, $\operatorname{index}(\bar{\partial}) = 1$.

For $M^2 = \mathbb{C}P_2$, $\chi(M) = 3$, $\operatorname{sign}(M) = 1$, $\operatorname{index}(\bar{\partial}) = 1$.

Thus

$$\left.\begin{aligned} 1 &= 4\alpha + 0\beta \\ 1 &= 3\alpha + 1\beta \end{aligned}\right\} \Rightarrow \alpha = \frac{1}{4}, \quad \beta = \frac{1}{4}$$

Here for $\mathbb{C}P_2(\text{-orn})$, $\operatorname{index}(\bar{\partial}, \text{-orn}) = \frac{3}{4} - \frac{1}{4} \notin \mathbf{Z}$.

Therefore $\mathbb{C}P_2(\text{-orn})$ is <u>not</u> a complex manifold.

Chapter VI

The Hirzebruch Signature Formula

Let M^m be an even dimensional compact orientable manifold and $*_p: C^\infty(\Lambda^p T^*M) \to C^\infty(\Lambda^{m-p} T^*M)$ the Hodge star operator. Recall that $*_{m-p}*_p = (-1)^{p(m-p)}$ and $d^* = (-1)^{mp+m+1}*d*$.

In the case where m is even, $*_{m-p}*_p = (-1)^p$ and $d^* = -*d*$.

Define $\tau_p = (-1)^{p(p-1)/2}*_p$.

Then

$$
\begin{aligned}
\tau_{m-p}\tau_p &= (-1)^{(m-p)(m-p-1)/2}(-1)^{p(p-1)/2}*_{m-p}*_p \\
&= (-1)^{(m^2-2mp-m+p^2+p+p^2-p)/2}(-1)^p \\
&= (-1)^{(m^2-m)/2+p^2-mp+p} \\
&= (-1)^{(m^2-m)/2}, \quad \text{since } m \text{ is even} \\
&= \begin{cases} 1 & \text{if } m \equiv 0 \bmod 4 \\ -1 & \text{if } m \equiv 2 \bmod 4 \end{cases}
\end{aligned}
$$

$$
\begin{aligned}
\tau_{p+1}d &= (-1)^{(p+1)p/2}*_{p+1}d = (-1)^{(p+1)p/2}(-1)^p*_{p+1}d*_{m-p}*_p \\
&= (-1)^{(p-1)p/2}*_{p+1}d*_{m-p}*_p = *_{p+1}d*_{m-p}\tau_p \\
&= -d^*\tau_p
\end{aligned}
$$

and

$$
\begin{aligned}
d\tau_p &= (-1)^{(p-1)p/2}d*_p = (-1)^{m-p+1}*_{p-1}*_{m-p+1}d*_p(-1)^{(p-1)p/2} \\
&= (-1)^{m-p+1}(-1)^{p(p-1)/2}(-1)^{(p-1)(p-2)/2}(-\tau_{p-1}d^*) \\
&= (-1)^{-p+1+p^2/2-p/2+p^2/2-3p/2+1}(-\tau_{p-1}d^*) \\
&=(-1)^{p^2-3p+2}(-\tau_{p-1}d^*) = (-1)^{(p-1)(p-2)}(-\tau_{p-1}d^*) = -\tau_{p-1}d^*
\end{aligned}
$$

Now let $\Lambda = \Lambda(T^*M) = \bigoplus_{p=0}^{m} \Lambda^p(T^*M)$ and $\tau: \Lambda \to \Lambda$ be defined by $\tau = \bigoplus_{p=0}^{m} \tau_p$.

Then the above results imply that

$$(d + d^*)\tau = -\tau(d + d^*) \quad \text{and} \quad \tau^2 = \pm 1 .$$

Decompose $\Lambda = \Lambda^+ \oplus \Lambda^-$ where

$$\Lambda^\pm = \begin{cases} \text{range } \frac{1\pm\tau}{2} & \text{when } m \equiv 0 \bmod 4 \\ \text{range } \frac{i\pm\tau}{2} & \text{when } m \equiv 2 \bmod 4 \end{cases}$$

Then $\Lambda^\pm$ are the $\pm 1(\pm i)$ eigenspaces of τ .
Since $d + d^*$ anticommutes with τ

$$d + d^*: C^\infty(\Lambda^+) \to C^\infty(\Lambda^-)$$

and

$$(d + d^*)^* = d + d^*: C^\infty(\Lambda^-) \to C^\infty(\Lambda^+)$$

The two step complex $d + d^*: C^\infty(\Lambda^+) \to C^\infty(\Lambda^-)$ is elliptic since $\frac{1}{\sqrt{-1}}\sigma(d + d^*)(\xi) = e(\xi) + i(\xi)$, which is non-singular ($e(\xi)$ denotes exterior and $i(\xi)$ interior multiplication by $\xi \in T^*M$) . This elliptic complex is called the <u>signature complex of M</u> and

$$\text{index}(d+d^*) = \dim(\ker(d + d^*): C^\infty(\Lambda^+) \to C^\infty(\Lambda^-)) - \dim(\ker(d+d^*): C^\infty(\Lambda^-) \to C^\infty(\Lambda^+))$$

is called the <u>signature of M</u>, denoted $\operatorname{sign}(M)$.

In fact $\operatorname{sign}(M)$ depends only on $\Lambda^{\pm} \cap \Lambda^{m/2}$. For example when $m \equiv 0 \bmod 4$, let ϕ be a harmonic p-form with $p \neq \frac{m}{2}$. Define $\phi^{+} \in C^{\infty}(\Lambda^{+} \cap \Lambda^{p})$ and $\phi^{-} \in C^{\infty}(\Lambda^{-} \cap \Lambda^{p})$ by $\phi^{+} = \frac{\phi+\tau\phi}{2}$ and $\phi^{-} = \frac{\phi-\tau\phi}{2}$. Then $\phi^{+} \neq 0 \neq \phi^{-}$ since $p \neq \frac{m}{2}$ and $\phi^{\pm} \in \ker(d+d^{*}): C^{\infty}(\Lambda^{\pm}) \to C^{\infty}(\Lambda^{\mp})$. Thus in computing $\operatorname{index}(d+d^{*})$ everything cancels except in the middle dimension.

When $m \equiv 0 \bmod 4$ we also have

$$\tau_{m/2} = (-1)^{m/2(m/2-1)} *_{m/2} = (-1)^{m^2/8} *_{m/2} = *_{m/2}.$$

Therefore $\operatorname{sign}(M)$ = multiplicity of $+1$ - multiplicity of -1 as eigenvalues of $*$: {harmonic $\frac{m}{2}$-forms} $\to$ {harmonic $\frac{m}{2}$-forms}. However, when $m \equiv 2 \bmod 4$

$$\begin{aligned}\operatorname{sign}(M) &= \text{multiplicity of } +i - \text{multiplicity of } -i \text{ as eigenvalues of } \tau_{m/2} \\ &= 0\end{aligned}$$

<u>Example</u>. $\operatorname{Sign}(\mathbb{C}P_{2n}) = 1$.

Let Ω be the Kaehler form on $\mathbb{C}P_{2n}$. It is well-known that $H^{2k}(\mathbb{C}P_{2n}, \mathbb{C})$ is generated by Ω^{k} and $H^{2k+1}(\mathbb{C}P_{2n}, \mathbb{C}) = 0$. $\mathbb{C}P_{2n}$ has real dimension $4n$ and

$$\dim\{\text{harmonic forms of degree } 2n\} = 1 .$$

Therefore sign($\mathbb{C}P_{2n}$) = ±1 .

One easily checks that $*\Omega^n = \Omega^n$, so that sign($\mathbb{C}P_{2n}$) = +1 .

The signature of a product manifold

Let M and N be compact oriented even dimensional manifolds and let $M \times N$ be equipped with the product orientation and metric. Then $\Lambda(M \times N) = \Lambda(M) \otimes \Lambda(N)$ and $\tau_{M\times N} = \tau_M \otimes \tau_N$ so that

$$\Lambda^+(M \times N) = \Lambda^+(M) \otimes \Lambda^+(N) \oplus \Lambda^-(M) \otimes \Lambda^-(N)$$

and

$$\Lambda^-(M \times N) = \Lambda^-(M) \otimes \Lambda^+(N) \oplus \Lambda^+(M) \otimes \Lambda^-(N)$$

from which it follows that $\text{sign}(M \times N) = \text{sign}(M) \cdot \text{sign}(N)$.

Thus, for example,

$$\text{sign}(\mathbb{C}P_{2r_1} \times \dots \ \mathbb{C}P_{2r_k}) = \text{sign}(\mathbb{C}P_{2r_1}) \dots \text{sign}(\mathbb{C}P_{2r_k}) = 1 .$$

The twisted signature complex

To motivate the definition of the twisted signature complex we first give an alternative definition of the ordinary signature complex.

The involution τ defined earlier anticommutes with both $i(\xi)$ and $e(\xi)$ for all $\xi \in T^*M$. Therefore

$$i(\xi) + e(\xi) \colon C^\infty(\Lambda^\pm) \to C^\infty(\Lambda^\mp) \quad \text{for all} \quad \xi .$$

Let D be the Riemannian connection on TM .

D naturally induces a connection on $\Lambda^\pm$ which we shall also denote by D . It is not hard to show that the following diagram is commutative.

$$\begin{array}{ccc} C^\infty(\Lambda^\pm) & \xrightarrow{\ d+d^*\ } & C^\infty(\Lambda^\mp) \\ & {\scriptstyle D}\searrow \qquad \nearrow{\scriptstyle i+e} & \\ & C^\infty(T^*M \otimes \Lambda^\pm) & \end{array}$$

Now let E be a C^∞ complex vector bundle over M with an Hermitian inner product. Let D_E be a Riemannian connection on E . The vector bundle $\Lambda \otimes E$ has a naturally induced inner product and Riemannian connection $D \otimes D_E$. Define $d_E \colon C^\infty(\Lambda^\pm \otimes E) \to C^\infty(\Lambda^\mp \otimes E)$ by making the following diagram commutative.

$$\begin{array}{ccc} C^\infty(\Lambda^\pm \otimes E) & \xrightarrow{\ d_E\ } & C^\infty(\Lambda^\mp \otimes E) \\ & {\scriptstyle D \otimes D_E} \searrow \quad \nearrow {\scriptstyle i+e} & \\ & C^\infty(T^*M \otimes \Lambda^\pm \times E) & \end{array}$$

d_E corresponds to $d + d^*$ of the ordinary signature complex and is a first-order self-adjoint elliptic operator. It is self-adjoint because $D \otimes D_E$ is a Riemannian connection and elliptic because the symbol of $D \otimes D_E$ is the identity. The elliptic complex $d_E : C^\infty(\Lambda^\pm \otimes E) \to C^\infty(\Lambda^\mp \otimes E)$ is called the <u>signature complex of M twisted by E</u>.

<u>The Index Theorem for the Twisted Signature Complex.</u>

Let $d_E^2 : C^\infty(\Lambda^+ \otimes E) \to C^\infty(\Lambda^+ \otimes E)$ be denoted by D^+ and

$d_E^2 : C^\infty(\Lambda^- \otimes E) \to C^\infty(\Lambda^- \otimes E)$ be denoted by D^-.

Let orn denote the chosen orientation. By the general index theory developed earlier we know that

$$\text{index}\,(d_E) = \int_M B_m(x, d_E, \text{orn})\,d\text{vol} \qquad \text{where}$$

$$B_m(x, d_E, \text{orn}) = B_m(x, D^+, \text{orn}) - B_m(x, D^-, \text{orn})$$

is a local invariant formula in the derivatives of the components of the metric tensor of M <u>and</u> the connection on E. Observe that

$$*_{orn} = -*_{-orn} \Rightarrow \tau_{orn} = -\tau_{-orn}$$

$$\Rightarrow \Lambda^{\pm}_{orn} = \Lambda^{\mp}_{-orn} \Rightarrow B_n(x,d_E,orn) = -B_n(x,d_E,-orn).$$

Thus $B_n(x,d_E,orn)$ is an invariantly defined N-form. Suppose E has fiber dimension r and $e_1,\ldots,e_r$ is a local frame for E. Let ω_{stk} be defined by

$$D_E(e_s) = \sum_{k=1}^{m} \sum_{t=1}^{r} \omega_{stk} e_t \otimes dx_k \quad ; \quad s = 1,\ldots,r.$$

Then $B_n(x,d_E,orn) = B_n(x,D^+,orn) - B_n(x,D^-,orn)$ is a skew-invariant polynomial homogeneous of degree n in the derivatives of g_{ij} and ω_{stk}

(where order $\omega_{stk/\omega'} = 1+|\omega'|$ for a multi-index ω').

The following theorem is the main step in the proof of the index theorem for the twisted signature complex. We only sketch the proof.

Theorem:

Let $P(g_{ij/\omega}, \omega_{stk/\omega'})$ be a skew-invariant homogeneous polynomial of degree $n \leqslant m$. i.e.

P: {metrics on M and connections on E} $\to$ N-forms on M and is invariant. Then

(a) $n < m \Rightarrow P \equiv 0$.

(b) $n = m \Rightarrow P$ is a combination of Pontryagin classes of TM and Chern classes of E.

Proof: (Sketch) A chart X and a local frame $\{e_s\}$ can be found such that $g_{ij/k}(x_0) = 0$ and $\omega_{stk}(x_0) = 0$.

Firstly, one can show that if P satisfies the conditions of the theorem then P has a monomial A of the form

$$A = g_{i_1 j_1/\omega_1} \cdots g_{i_\ell j_\ell/\omega_\ell} \cdot \omega_{s_{\ell+1} t_{\ell+1} k_{\ell+1}/\omega'_{\ell+1}} \cdots \omega_{s_r t_r k_r/\omega'_r}$$

such that

$$\deg_a(g_{i_1 j_1/\omega_1} \cdots g_{i_\ell j_\ell/\omega_\ell}) = 0 \quad \text{for} \quad a > 2\ell$$

and

$$\deg_a(\omega_{s_{\ell+1} t_{\ell+1} k_{\ell+1}/\omega'_{\ell+1}} \cdots \omega_{s_r t_r k_r/\omega'_r}) = \begin{cases} 0 & \text{if} \quad a < 2\ell \\ \text{at least } 1 & \text{if} \quad a \geqslant 2\ell + 1 \end{cases}$$

Using the fact that $|\omega_i| \geqslant 2$ for $1 \leqslant i \leqslant \ell$ and $|\omega'_i| \geqslant 1$ for $i \geqslant \ell + 1$ one can show by a counting argument that $r = \frac{m}{2}$ so that $P \equiv 0$ for $n < m$, which proves (a), and that if $n = m$, P must have a monomial of the form

$$A = g_{i_1 j_1/k_1 \ell_1} \cdots g_{i_r j_r/k_r \ell_r} \cdot \omega_{s_{r+1} t_{r+1} u_{r+1}/v_1} \cdots \omega_{s_{\frac{m}{2}} t_{\frac{m}{2}} u_{\frac{m}{2}}/v_{\frac{m}{2}}}$$

where $g_{i_1j_1/k_1\ell_1}\cdots g_{i_rj_r/k_r\ell_r}$ involves only the indices $1,\ldots,2r$ and $\omega_{s_{r+1}t_{r+1}u_{r+1}/v_{r+1}}\cdots\omega_{s_{\frac{m}{2}}t_{\frac{m}{2}}u_{\frac{m}{2}}/v_{\frac{m}{2}}}$ involves only the indices $2r+1,\ldots,m$.

This fact can be used to show that there is a 1-1 map from $\mathcal{R}_m$ into $\sum_{p=0}^{m}(\mathcal{P}_p \times \mathcal{Q}_{m-p})$ where

$\mathcal{R}_m$ = the space of skew-invariant polynomials of order m in $g_{ij/\omega}, \omega_{stk/\omega'}$

$\mathcal{P}_i$ = the space of skew-invariant polynomials of order i in $g_{ij/\omega}$

$\mathcal{Q}_i$ = the space of skew-invariant polynomials of order i in $\omega_{stk/\omega'}$

from which we obtain $\dim \mathcal{R}_m \leq \sum_{p=0}^{m} \dim \mathcal{P}_p \dim \mathcal{Q}_{m-p}$. Now there are exactly $\sum_{k=0}^{m/4} \pi(k)\pi_r\left(\frac{m-4k}{2}\right)$ combinations of Pontryagin classes of TM and Chern classes of E of total order m (here $\pi_r(j)$ = the number of partitions of j into integers $\leq r$, r = fiber dimension of E), and such combinations are clearly skew-invariant polynomials of degree m in the $g_{ij/\omega}$ and $\omega_{stk/\omega'}$. Thus it is sufficient to show that

$$\dim \mathcal{P}_p \begin{cases} = 0 & \text{when } p \not\equiv 0 \mod 4 \\ \leqslant \pi\left(\frac{p}{4}\right) & \text{when } p \equiv 0 \mod 4 \end{cases}$$

and

$$\dim \mathcal{Q}_p \begin{cases} = 0 & \text{when } p \not\equiv 0 \mod 2 \\ \leqslant \pi_r\left(\frac{p}{2}\right) & \text{when } p \equiv 0 \mod 2, \end{cases}$$

since this would imply $\dim \mathcal{R}_m \leqslant \sum_k \pi(k)\pi_r\left(\frac{m-4k}{2}\right)$ which in turn would imply that each element of $\mathcal{R}_m$ is a combination of Pontryagin classes and Chern classes.

So let $P \in \mathcal{P}_p$. One can show that P has a monomial

$$A = g_{11/i_1 x_1} \cdots g_{p/2\ p/2/i_{p/2} x_{p/2}}$$

$$\{i_1, \ldots, i_{\frac{p}{2}}\} \in \mathrm{Perm}\left(\frac{p}{2}\right) \quad \text{and} \quad x_1, \ldots, x_{\frac{p}{2}} \geqslant \frac{p}{2} + 1.$$

Suppose $\tau \in \mathrm{Perm}\left(\frac{p}{2}\right)$ is given by $\tau(j) = i_j$. Then it follows that $\dim \mathcal{P}_p \leqslant$ number of distinct conjugacy classes of such $\tau = \pi\left(\frac{p}{2}\right)$. In fact we can find A for each $P \in \mathcal{P}_p$ such that the lengths of all cycles of τ are even, so that $\dim \mathcal{P}_p \leqslant \pi\left(\frac{p}{4}\right)$.

To prove that $\dim \mathcal{Q}_p \leqslant \pi_r\left(\frac{p}{2}\right)$ one shows that there is an injection $\mathcal{Q}_p \to Q^I_{\frac{p}{2}}$ where

$$Q^I_{\frac{p}{2}} = \{\text{homogeneous invariant polynomials of degree } \tfrac{p}{2} \text{ in } \omega_{st}\}$$

with $\omega_{st} = \sum_{j,k=1}^{m} \omega_{stj'k} dx_j dx_k$.

One can then argue that $\dim \mathbb{Q}^{I}_{\frac{p}{2}} \leq \pi_r\left(\frac{p}{2}\right)$.

Corollary (Index Theorem for Twisted Signature Complex.)

(1) $B_n(x,d_E,\text{orn}) = 0$ for $n < m$

(2) $\int_M B_m(x,d_E,\text{orn})\,d\text{vol} = \text{index}\,(d_E) =$ signature of M with coefficients in E,

where $B_m(x,d_E,\text{orn})$ is a combination of Pontryagin classes of TM and Chern classes of E.

In the special case where E is the trivial bundle,

$$d_E = d+d^*: C^\infty(\Lambda^\pm) \to C^\infty(\Lambda^\mp)$$

and we get the index theorem for the ordinary signature complex:

From the previous corollary we know that $B_m(x,d+d^*,\text{orn})$ is some Pontryagin class of TM Suppose $\dim M = m = 4k$. Let $r = (r_1,\ldots,r_k)$ be a multi-index of nonnegative integers such that $\sum_{i=1}^{k} r_i = k$. Define the $4k$ (real) dimensional manifold

$$M_r = \mathbb{C}P_{2r_1} \times \ldots \times \mathbb{C}P_{2r_k}.$$

The total number of such M_r is exactly $\pi(k)$. It can be shown that the M_r form a dual basis to the Pontryagin classes of TM, i.e. if P is a Pontryagin class such that $\int_{M_r} P = 0$ for all M_r then $P \equiv 0$.

Now $\text{sign}(\mathbb{C}P_{2r_i}) = 1 \Rightarrow \text{sign}(M_r) = 1$.

The Hirzebruch L_k polynomial in the Pontryagin classes of TM is defined by the relations $\int_{M_r} L_k \, dvol = 1$. Therefore

$$\int_{M_r} L_k \, dvol = \text{sign}(M_r) = \int_{M_r} B_m(x,d+d^*,orn)dvol$$

$$\Rightarrow \int_{M_r} (L_k - B_m)dvol = 0 \quad \text{for all } M_r \Rightarrow L_k \equiv B_m.$$

<u>Corollary</u>: Index $(d+d^*) = \int_M L_k \, dvol$.

REFERENCES

1. M. Atiyah, R. Bott , The Index Problem for Manifolds with Boundary Differential Analysis (Bombay Colloquium) Oxford 1964

2. M.Atiyah,R. Bott,and V.K.Patodi, On the Heat Equation and the Index Theorem Inventiones Mathematica, 19 Fasc 4 1973

3. M. Atiyah, R. Bott, and A. Shapiro, Clifford Modules Topology Suppl 1 3 1964 3-38

4. M. Atiyah, I.Singer, The Index of Elliptic Operators III Ann. Math 87 1968, 546-604

5. M.Atiyah, V.K.Patodi and I.M.Singer, Spectral Asymmetry and Riemannian Geometry, Bull London Math Soc, 5 1973 ,229-234

6. S.Chern, A simple intrinsic proof of the Gauss-Bonnet formula for closed Riemannian Manifolds, Annals of Math 45 1944,747-752

7. S.Chern, On the curvatura integra in a Riemannian manifold Annals of Math 46 1945 ,674-684

8. P.Gilkey, Curvature and the Eigenvalues of the Laplacian for Elliptic Complexes, Advances in Math 10 1973, 344-382

9. P.Gilkey, Curvature and the Eigenvalues of the Dolbeault Complex for Kaehler manifolds, Advances in Math 11 1973,311-325

10. P.Gilkey, The Boundary Integrand in the formula for the Signature and Euler Characteristic of a Riemannian manifold with boundary, to appear in Advances

11. P.Greiner, An asymptotic expansion for the Heat equation Proc Symp Global Analysis 10, 1968 AMS, 137-146

12. L.Hormander, Pseudo-differential Operators and Non-elliptic Boundary Problems, Annals of Math 83, 1966 , 129-209

13. H.P. McKean and I.M.Singer, Curvature and the eigenvalues of the Laplacian, J.Diff. Geo,1 1967 43-69

14. V.K.Patodi, Curvature and the eigenforms of the Laplace operator, J. Diff. Geo 5 (1971),233-249

15. V.K.Patodi , An analytic proof of the Riemann-Roch-Hirzebruch theorem for Kaehler manifolds, J. Diff. Geo 5 1971, 251-283

16. R.T.Seeley, Complex powers of an elliptic operator Amer. Math. Soc. Proc. Symp. Pure Math 10,1967 ,288-307

17. R.T.Seeley, Topics in Pseudo-differential Operators CIME 1968, 167-305

18. H. Weyl, "The Classical Groups" Princeton University Press, Princeton N.J. 1946